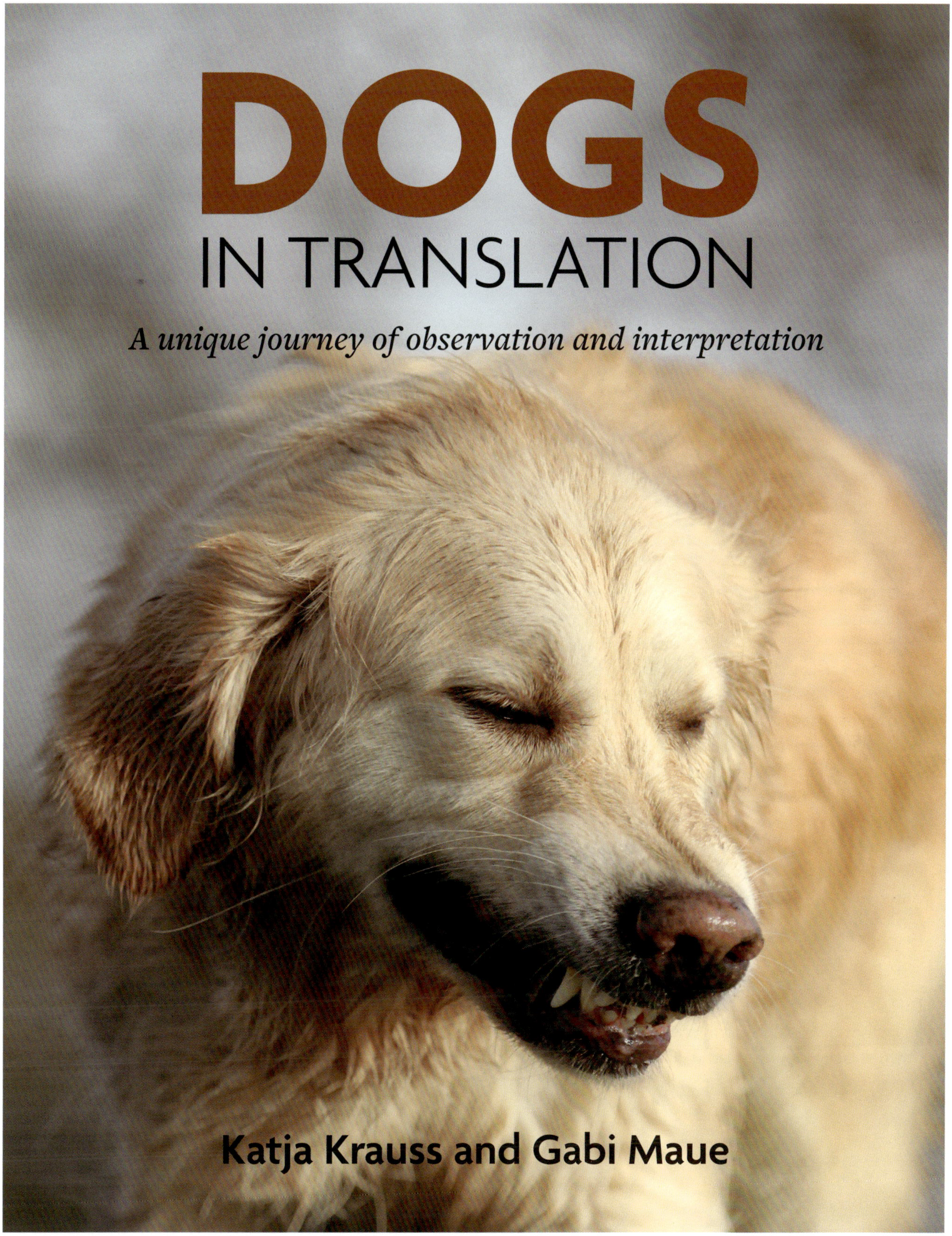
DOGS
IN TRANSLATION
A unique journey of observation and interpretation
Katja Krauss and Gabi Maue

ACKNOWLEDGEMENTS

This book never could have become a reality without the co-operation of the many dogs and people to whom we are so very grateful. We would like to thank our own dogs, who were the stars of this production, and all the many other dogs who are featured in our book.

In addition, we would like to thank:
Lotte Maue who helped us to bring some logic and structure into the first draft, and Andreas Krauss for support in transferring the photos and text to digital form. Udo Kopernik of the German Kennel Club (VDH), who made the first draft layout the book, allowing us to see what our dream would look like.
Gisela Rau, Nicole Hilgers and Erika Dick from Kynos-Verlag, our German publisher.
A special thank you goes to Her Highness Sheikha Latifa Bint Mohammed bin Rashid al Maktoum. She and her family made it possible to capture an impression of the love towards dogs in the United Arab Emirates.
We owe a very special heartfelt thanks the wonderful Linda Tellington-Jones for her inspiration and for the Foreword she has contributed. Also thanks to our Tellington colleagues: Karin-Petra Freiling, Lily Merklin und Dr. med. vet. Daniela Zurr who shared their expertise in special articles in this book.
We thank Kristina Maue-Limbach for her help with Linda's Foreword and Christine Schwartz for her quick and diligent translation of our work into English. Thank you to Martina Braun, Romana Hallada and Kathy Zozula for proofreading.
To all of you who have allowed us to use your pictures, we own a huge debt of gratitude.
Our most special thanks goes to the families of the authors who patiently stood by and made it possible to spend years on this very special project.

PHOTOGRAPHY
All photos, including cover by Krauss, Katja and Maue, Gabi, except: Betz, Ingrid: p. 98; Boiselle, Gabriele (www.editionboiselle.de): P.6/7; Braun, Ursula: P. 365 t. le., 441 all; Büschel, Marion: P. 253, 255 t.; Grobecker, Lina: P. 613 b.; Grote, Esther: P. 225 r.; Hares, Michaela: P. 8 t., 10 b., 139 b., 215 t.; Hoffmann, Natascha: P. 8 mi.; Hoffmann, Sandra: P. 534 mi.; Janssen, Anne: P. 18, 259 t., 585 b.; Kehrl, Jan: P. 178/179; Kess, Volker: P. 9 b.; Kiefer, Anja: P. 9 mi.; Kopernik, Renate: P. 8 b.; Kovanidis, Pantelis: P. 9 t.; Kuhn, Annalena: P. 40 ri. b., 44/45, 153 b., 462 Photo 2+3, 504 mi. b., 505 mi. b., 506/507, 569 b., 598 – 601, 608 mi., 618 b. ri.; Kuhn, Sabine: P. 257 b.; Löffler, Silke: P. 271 b. ri.; Maue-Limbach, Kristina: P. 88 b., 122 b., 249 all, 317 both t. 380 b., 422 t., 492, 493, 607 t, Maue, Lotte: P. 380 b.; Maue Wolfgang: P. 20; Mehrwald, Judith: P. 139 t.; Meisel, Thomas: P. 11 t.; Meissner, Jutta: P. 323 t.; Munninger, Nicole: P: 270 t. ri, 370, 378 all, 384/385, 489 all, 502 all four ; Swiatkowska, Marta: P. 610 both top; Pecqueux, Mathieu: P. 604/605; Raidt, Peter: P. 320 b. li.; Rau, Gisela: P. 250/251; Rickert, Dr. Daniela: P.10 t.; Schöning, Dr. Barbara: P. 10 m.; Sieber, Antonia: P. 78/79, 245 t., 270 t. le., 276 m, 552 b. ri.; Wild, Ingrid: P. 254 t. le., 266 all, 267 all.

First published in Germany in 2020 by Kynos Verlag

English language edition published in 2022 by First Stone Publishing,
an imprint of Westline Publishing Limited
The Old Hen House, St Martin's Farm, Zeals, BA12 6NZ, United Kingdom

Designed by Alan Cooper

ISBN 9781910488645

Printed by Printworks Global Ltd., London & Hong Kong.

1234567890

CONTENTS

I have known both authors, Katja Krauss and Gabi Maue, since their training as Tellington TTouch® practitioners and truly appreciate their work and deep knowledge of dogs.

In 2014 I first saw the first draft for this book, which has now, some years later, become reality. At that time, I was already very impressed by the scope of the subject matter, the selection and the quality of the photos, and the precise photo descriptions. It raised the hope for a completely new way of discussing and explaining the non-verbal communication of dogs. It allowed the complexity of the topic to be made visible and understandable, not only for people who deal with canine behaviour professionally, but for everyone on the other end of the leash.

Such a book has been urgently needed for some time, and I am very happy that it is now finally finished.

This book gives new and unprecedented access to the language of dogs and completes, and enriches, the selection of books on the body language and the behaviour of dogs. It contains a phenomenal collection of photos showing different breeds in numerous situations with detailed descriptions of their expressive behaviour. Each page invites you to dive deeper into the world of non-verbal communication between dogs.

Without exaggeration, I would speak of this as an epic work. It should be translated into many languages and be made accessible to a large circle of people who love dogs!

For our Practitioner Training for Dogs, I hope that this book will become compulsory reading! It is a wonderful, indispensable addition to the philosophy (and working method) of Tellington TTouch®.

Linda Tellington-Jones

PREFACE

We, the authors Katja Krauss and Gabi Maue, have been working with the Tellington TTouch® method for many years, applying it to dogs of a variety of breeds, ages, characters and life experiences. Many people who have come to us clearly love their dogs, and think they understand them. But we noticed, time and again, that they either didn't pick up on their dog's emotions or completely misunderstood them.

In many instances, this stemmed from cultural preconceptions. For example, an individual may be influenced by how their parents dealt with the family dog; outdated training methods, based on punishment, may have been used, or there may be a prevailing fear that the dog will 'dominate' the family. Inevitably, these perceptions lead to a misinterpretation of body language and signalling.

We did a lot of research and found that although there are some books on the market that describe communication through body language, they all lack detailed descriptions through photos. We were determined to fill this gap and so we started taking photos of dogs in all situations: when they were interacting with people, with other dogs, or when they were just doing their own thing. Of course, we were aware that a photo can only capture a moment in time, and this can be difficult to interpret. We, therefore, decided to include a number of sequences where we took around seven photos per second.

The communication between dogs happens incredibly fast. In most cases it is, initially, very subtle before the dog resorts to stronger signals that cannot be missed. This subtle communication is usually so minute that we don't even see it. It changes within a split second and, depending on the situation, it can be different from one side of the body to the other – for example, when a dog finds himself between two other dogs, who are both sending different signals.

Very few people are capable of catching this quick progression – and not all dogs are capable of it either. Each individual is different and has a unique set of skills. When we were sorting through the photos, we noticed that we sometimes relied on a gut feeling when evaluating a situation, especially just before an escalation of behaviour.

Short-haired and light-coloured dogs are easier to read than those with long hair, or who are a dark colour. Dogs with markings on their faces are often easier for other dogs to read than those with a solid coloured face. However, face markings can sometimes become a problem for dogs who don't like to be stared at, because the marking around the eyes can give the impression that the dog is fixating.

We decided to use our own dogs in many of the photos where we show different emotions and moods, because we know them best. We have a range of breeds, which is useful for interpreting the differences in body language between breeds and breed types. However, what we did learn is that every dog is an individual. Each dog shows his feelings with a different level of intensity, and prefers some signals over others. Depending on their character and temperament, a dog will display body language that is either guarded and restrained, or bold and confrontational.

As a result of these differences, we often evaluate these signals very differently.

Take the example of Quivive, a Pyrenean Sheepdog, who is owned by Gabi. She commented that her dog looked really bored. However someone, who has several dogs of the same breed, said that that if that was her dog in the photo, she would anticipate him taking off in an instant.

This was a head study of Quivive, so the interpretation was based solely on her facial expression. It would have been helpful to see her whole body to determine her emotional state, and then compare this with the dog who was, maybe, planning to take off. This example illustrates why we don't just concentrate on the dog's face, but must also consider the entire body.

It is also important to consider external factors before attempting to evaluate a dog's emotional state. For example, a dog's expression and body language can be influenced by the setting (inside or outside), the weather, the temperature, or whether he is in light or shade. We should also consider internal factors, such as hunger and thirst, or whether the dog is in discomfort or pain.

All these facts relate to each other and have an impact, which is why we start the book with detailed observations of individual body parts before we progress to connecting the dots. Our aim is to train the reader's eye and build awareness of a dog's emotional state, whether this is their own dog, or any other dog they may meet. In time, this will develop that gut feeling we discussed earlier. In situations where a dog is meeting people, or other dogs, he may need support, depending on his comfort level and experience. We, therefore, need to work out when to step in, and what form of assistance is required. It is important that the dog understands our intentions so he does feel under greater pressure, which would have an adverse effect on our relationship.

Dogs feel emotions and display them. Joy, anger, fear, curiosity, love, revulsion, and disappointment – these are all emotions that we think we have seen in our dogs. However, we are aware that different people have very different perceptions of how a dog may be feeling. How can we truly recognise and interpret the emotions of another species? Cynology, the study of matters relating to domesticated dogs, is still in its infancy when it comes to researching emotions, and the terminology that is used to express how a dog is feeling. For example, researchers are divided as to whether dogs experience jealousy or remorse. Could it be that dogs have as many different sensations for the emotion of 'joy' as the multitude of words the Inuit have to describe 'snow' (John Bradshaw *Dog Sense*, 2011).

Does one dog use different body language and signals to express joy in comparison with another? Is a dog 'jealous' in the way we interpret the word, or does it come from a different emotion that causes him to respond in a way that simply reminds us of jealousy? Could it be connected to separation anxiety, guarding, defending resources, or feelings of concerns that makes him want to keep other dogs, or people, out of his space?

Our goal is to help you understand your own dog, picking up on his subtle signals and body clues, which communicate how he is feeling. We recommend a detailed study of the photos in this book, and then suggest you take numerous pictures of your own dog in a wide range of situations. Examine these photos carefully and practise observing them with a neutral eye. Observe and describe what you see before you try to interpret.

The first part of the book focuses on a detailed examination of each the body parts that are used to communicate. We then examine the entire body, and the specific situation, to interpret the dog's emotional state.

We are of the opinion that nobody can own another living being and, for this reason, we have used the word 'guardian', rather than 'owner' because it expresses what we would like to be for our animals: responsible protectors and guardians of their welfare.

We all want to give our dogs the best possible life. It is crucial that they are loved but, it is equally important that we recognise their needs and fulfill them. It is our duty to bring respect and understanding into their daily lives, and every situation we present them with. We hope that our book will make this possible.

PART ONE:

WHAT DOES THE DOG'S BODY TELL US?

SECTION I:

FOCUS ON THE HEAD

1. POINTS OF THE HEAD

When we look at the dog's head we can break it into different sections to gather detailed information, and also evaluate it as a whole. These are the important parts of the head that will help us interpret a dog's emotional state and his state of mind.

1. The ears and the base of the ears.
2. The top and midline of the head.
3. Eyes.
4. Mouth and corner of the mouth.
5. Nose/nostrils.
6. Cheeks.
7. Bridge of the nose.

2. FACIAL EXPRESSIONS

A dog's facial muscles are very complex. There are numerous small strands of muscle in the connective tissue, which lies just under the skin, that influence facial expression. Depending on the situation, expressions can be asymmetrical. Facial expression is an important form of communication and, with the correct interpretation, we can learn a great deal about a dog's state of mind.

Note the subtle raising of the corner of the mouth. This is on the left side only and is a response to a dog approaching from that side.

Here there is clear evidence of a raised upper lip, wrinkles on the side of the nose, and visible canines.

The left eyebrow is raised and there is wrinkling above the eye.

This puppy is blinking asymmetrically due to a caring mother on the same side.

ONE DOG – TWO FACES

Much like people, dogs can have asymmetrical faces. This can be exaggerated by distinctive facial markings or uneven ears, each having a potentially different effect (*see The eyes, page 45*). We have a tendency to focus more on the optically emphasised half of the dog's face.

A relaxed, happy face with two almost identical sides; the angle of the mouth is also mirrored. The overall impression of this dog is very similar on both sides.

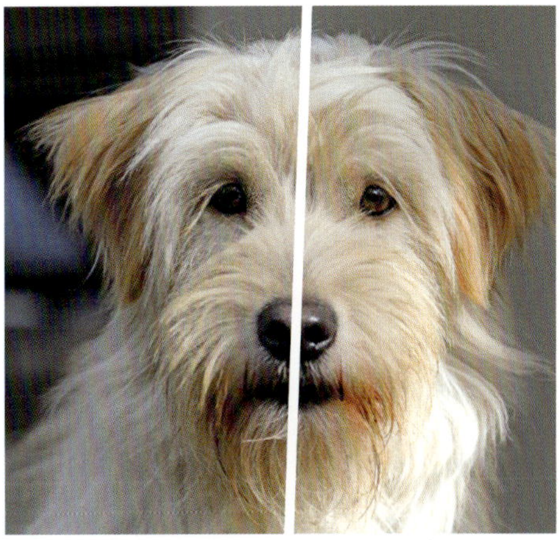

A relaxed face with eyes that seem to be different sizes. In this photo, the larger eye (*left*) seems to be focused on something in particular.

A slightly tense, alert face with asymmetrical markings and different ear positions. The impression of direct eye contact is reinforced by the markings around the right eye, the position of the ear, and the apparent wrinkling on top of the head.

An alert face, which shows some tension. This dog has extreme asymmetrical markings and has a slightly slit-shaped eye. With this dog, the direct eye contact – accompanied by his dark eye, dark fur and a light speck in the eye – can be confusing to other dogs.

Looking at the photos which show different facial expressions on the same dog, how do you respond? Try covering one half of the dog's face and studying his expression, and then repeat the experiment, covering the other side. What do you notice? Do your perceptions differ depending on what side of the face you are looking at? What feelings do the different facial expressions invoke in you?

3. WRINKLES

Wrinkles on the top of the head and forehead give us clues about a dog's mood and his state of mind. Ear carriage, and the positioning of the eyebrows, add to the development of the wrinkles. The depth and amount of wrinkling can tell us if the dog is attentively listening or relaxed. Facial wrinkles, and those on the forehead, also convey whether a dog is stressed, fearful, angry or annoyed. Mostly, the changes in the wrinkles are very subtle, except when a dog is threatening, in which case we would also see wrinkling around his mouth and nose.

When looking at wrinkles, it is important to be aware of how many wrinkles are normal for that individual dog, and the depth of the wrinkles when he is relaxed. To be able to respond proactively, it is helpful to identify situations when the dog starts to show wrinkles that change his facial expression. Relaxed dogs usually show very few wrinkles on their face, with the exception of dogs that have wrinkly faces in their breed profile. Breeds that fall into this category, such as the Bloodhound and the Shar Pei, are often misread and misunderstood, especially by dogs without experience of wrinkly-faced breeds.

Wrinkles are easier to recognise in short-haired dogs than those with long facial fur coverings.

FROM TENSE TO RELAXED

Bonnie is a little American Staffordshire female with a high energy level and a lot of muscular tension. This is clearly evident by the degree of facial tension shown here. The skin on the top of her head is tight, due to pulling her ears back when she is tense.

After being led through the elements of the Tellington Confidence course, she becomes a lot more relaxed. The skin on top of her head softens, and the boney ridge is more visible, as she relaxes.

FROM RELAXED TO TENSE

Nala, a female Boxer cross, and her guardian, are taking part in a training course. They are watching the activities in the clinic with great interest. Initially, Nala (*left*) is relaxed which can be seen from the smooth outline of her forehead. As she becomes more focused and alert (*right*), her increased muscular tension produces wrinkles on the top of her head, and forehead. The wrinkles are created when she lifts her ears and furrows her eyebrows. This also changes the size of her eyes.

RELAXATION THROUGH CONTACT

Here Quivive, a Pyrenean Sheepdog, meets a female Beagle who clearly shows wrinkles on top of her head. She is interested in making contact with Quivive, but keeps her distance and stretches her neck and head. She becomes more relaxed as she makes contact.

The Beagle is clearly more relaxed; there are fewer wrinkles on her forehead, and they are flatter. Both dogs are encouraging more contact.

SLIGHT MUSCULAR TENSION TRIGGERED BY INTEREST

The light-coloured male Labrador meets the black dog, Taschka, and light-coloured Tamika (*in the foreground*). He is interested in Tamika and is excited. The hairs on his back are standing up. His raised ears cause wrinkles to form on his head.

Here we can see that the wrinkles on the Labrador's head only appear on the side that is facing the dogs. Tamika's ears show the typical 'flirting' position; they are clearly pointing to the side and back. This ear position is used when there is interest in making contact between a male and female.

MONITORING MUSCULAR TENSION

This male Sloughi lives in Tunisia on a large breeding facility. He is used to having contact with his handlers, but rarely meets strangers. In the following photos, he is walking past the photographer. She did not change her position or posture as he passed her.

Slight facial tension

Here we can see sight facial tension – a wrinkling of the brows – as the dog sights the photographer. Note the positioning of his ears:

- Base of the ears are slightly raised
- Ears carried towards the back
- The tips of the ears hang down

Increased facial tension

As he gets closer to the photographer there are more wrinkles on his forehead between the ears.

- Base of the ears are more raised
- Ears are carried further forward, the outside of the ears point forward
- The tips of the ears are held tighter

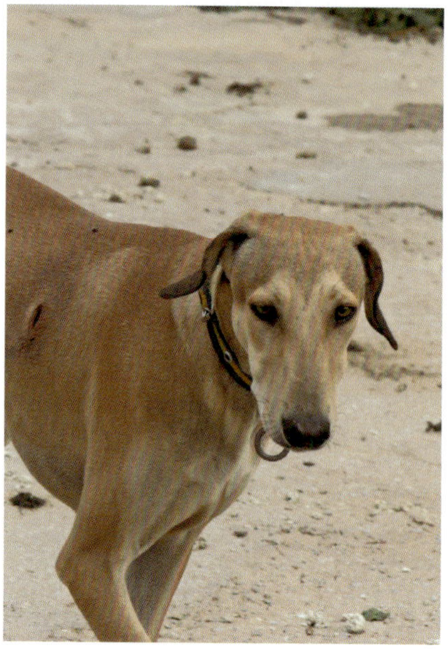

Releasing facial tension

The wrinkles have disappeared but the ear carriage – held towards the back – shows that he is still concerned about the stranger.

- Base of the ears are raised less and point towards the side and back
- Ears are held flat against the head and carried towards the back

CHANGING PERSPECTIVE

This series of photos show a different point of view regarding the wrinkling on the top of the head in relation to ear carriage. The ear position depends on the degree of relaxation.

This is the viewpoint when the dog is on leash and walking ahead of his handler. Generally it is not as easy to observe the wrinkles on the dog's face from this angle. However, we can clearly notice the position of the ears, which tells us the degree of focus or unease the dog is experiencing, as well as where he is looking. Many dogs will signal early on that they have seen, heard or scented something in the distance by raising the base of their ears and showing wrinkles. This gives us more time to respond proactively, especially if the dog has insecurities with regard to meeting other dogs or people on a walk, or if he has a strong prey drive.

Relaxed

- The top of the head is smooth
- The ears are relaxed and held towards the side and back

Tense

- Wrinkles on top of the head
- Ears raised at the base

With long-haired dogs, is not possible to see the changes on the top of the head but positioning of the ears give clear evidence of the degree of tension.

Slight muscular tension

- The ears are raised at the base and are held sideways
- The tips of the ears are relaxed and flop down

From this, we can conclude that the top of the head is still smooth and relaxed.

Increased muscular tension

- The ears are turned towards the back at the base
- The tips of the ears are pointing downward

Since the tips of the ears are closer together, we can assume there is an increase in tension and the top of the head is wrinkled.

The changes in this dog are very subtle and not easy to spot. However, if you observe the ear position of a long-haired dog closely, you will learn to identify these small changes and be able to evaluate them.

Strong muscular tension

- The ears are pointing towards the back at the base.
- The tips of the ears are pointing back and down and are held away from the body.

We can assume that the overall body tension has increased and the top of the head has deeper wrinkles and is even more tense.

UNINTENTIONAL CHANGES

There are a number of breeds that are heavily wrinkled, which other dogs find hard to read, especially if they have no experience with these breeds. Dogs with wrinkly faces can seem like they are staring, or threatening, even though they are just expressing interest in something they see.

Dogue de Bordeaux

Bulldog

Rhodesian Ridgeback cross

MISUNDERSTANDINGS DUE TO BREED DIFFERENCES

A male French Bulldog stands between a group of curious dogs. Looking at the numerous wrinkles above his eyes we would initially assume he is in alert mode. However his ears are in more of a 'flirting' position. This presents a confusing picture, and illustrates why it is important to observe and evaluate the signals from all the body parts.

Jean, the light-coloured dog is clearly impressed by the facial expression of the French Bulldog, notably his black facial markings (mask) and his facial wrinkles. She shifts her point of balance back (*left*), and is undecided as to whether she should initiate play or run away. In the photo (*right*) we see she opts for the latter.

CHANGE OF EXPRESSION DUE TO A MUZZLE OR HEAD HALTER

Wearing a muzzle or head halter can inhibit facial expressions. It becomes difficult for the dog to move his fine facial muscles. This can still happen when the muzzle or head halter is worn without pressure from the leash, even though there are now some very soft, padded models on the market. This restriction is heightened when the muzzle or head halter is not fitted correctly, or if the dog feels uncomfortable wearing this type of equipment. He will show his displeasure on his face, but his sensation of discomfort is not restricted to the part of the body that is restricted; it affects the entire dog.

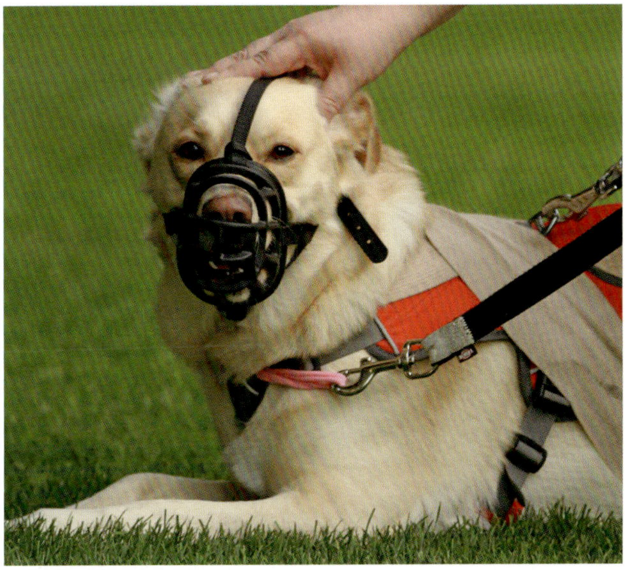

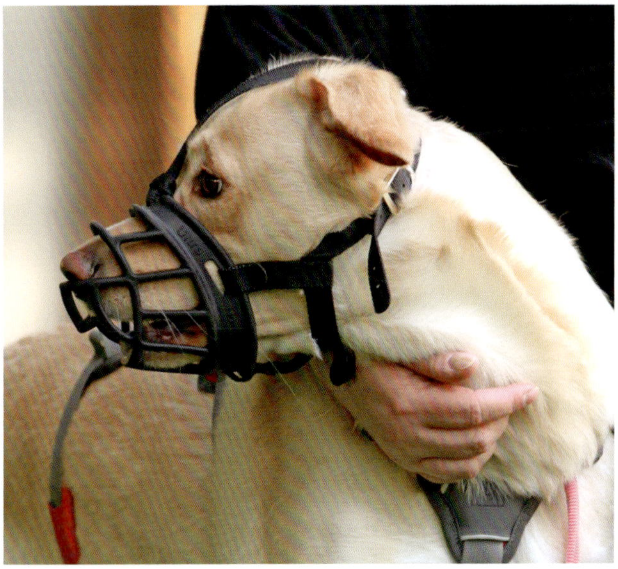

Here we can see how a muzzle changes the dog's facial expression because some parts of his face (his eyes) are pulled upwards, and other parts (his lips and the corner of his mouth) are difficult to see. This can limit communication with other dogs.

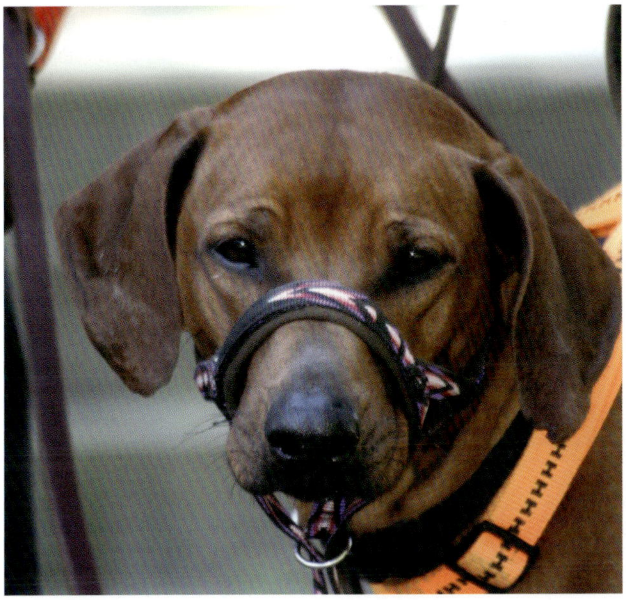

A head halter can change a dog's facial expression when the corner of his mouth is blocked and difficult to see.

Head halters and muzzles need to be wide enough around the mouth, and nose, so the dog can pant to regulate his temperature. It is important that he can open his mouth for clear communication, and also to calm himself.

RECOGNISING EMOTIONS

Since 1978 there has been considerable interest in standardising and describing facial mimicry (amongst others, Paul Ekman, Wallace V. Friesen: *Facial Action Coding System: A Technique for the Measurement of Facial Movement*. Consulting Psychologists Press, Palo Alto, Calif. 1978, OCLC 605256401). FACS is a technique that has been devised to recognise emotions through facial mimicry in humans, and the studies have been adapted to different animal species, including dogs. (DogFACS, 2013).

The meaning of the different facial expressions seems to be universal, but can vary in different breeds, which make it difficult for dogs to interpret the signals of another breed. Therefore it is important that a dog is not just raised with friends of his own type, but is also exposed to a wide variety of different breeds and types of dogs during his first few months so he can learn the different 'dialects' of facial expressions.

Did you know that the different ways dogs use their facial mimicry/expressions also influences humans? Over the past thousands of years, dogs have developed a muscle that enables them to pull the inner eyebrow upward. Wolves don't have this small muscle. When dogs pull their eyebrows upwards their eyes appear larger and therefore more youthful. This expression is similar to a human who is feeling sad. Researchers have concluded that dogs trigger this reflex in humans, prompting us to take care of them. In a study developed by DogFACS, research revealed that rescued dogs, who raised their eyebrows more frequently, were adopted more quickly than those who did not. *(Source: Wikipedia, Bridget M. Waller, Kate Peirce, Cátia C. Caeiro, Linda Scheider, Anne M. Burrows, Sandra McCune, Juliane Kaminski, Claire Wade: Paedomorphic Facial Expressions Give Dogs a Selective Advantage. In: PLoS ONE. 8, 2013, S. e82686, doi:10.1371/journal.pone.0082686).*

Who doesn't know this devoted look on a dog's face that melts everyone's heart?

4. HEAD CARRIAGE AND NECK

A dog will carry his head at different heights in certain emotional situations, so the neck, therefore, affects his ability to express his feelings. The height at which the head is carried is dependent on the engagement of the neck – raised or lowered – and gives us valuable information as to whether the dog is relaxed or alert.

We will keep reminding you to observe the head carriage and neck throughout this book. A relaxed dog will usually carry his head level with his back or lower, while a dog experiencing tension will raise his head. The exceptions are those breeds which naturally carry their heads higher. This includes many of the Nordic breeds, the Spitz breeds, and the Japanese Akita.

There are also situations (e.g., while crouching) that are connected to body tension and are indicated by a lowered head carriage. Carrying the head lower than normal can also be seen in older dogs as they start to lose mobility in the vertebrae of their neck, or if they are ill or in pain. Dogs that feel insecure in a certain situation will often show a lower head carriage. Remember, when observing the head carriage and neck, it is important to look at the overall picture.

Changing the position of the head (raising or lowering) can also indicate a change in emotions. The best way of identifying whether a dog has raised or lowered his head is to observe where his nose points.

CHECK OUT WHERE THE NOSE IS POINTING

Dhanyi, a Golden Retriever, has noticed something in the distance. While changing her focus she raises her head slightly. The change of head carriage (noticeable by the light shadow in the background), a more open eye, and the direction she is looking, is minimal, but observable.

Dhanyi is not fixated on a specific object. She is looking down slightly.

Minimal facial tension

- Head raised
- Eye slightly almond shaped
- Upper eyelid with eyelashes lowered
- Eyebrows neutral
- Nose slightly pointing downwards

Dhanyi's focus is higher up and she is looking into the distance.

Increased facial tension

- Neck raised
- Eye opened wider
- Eyebrow lifted slightly
- Head raised slightly

HEIGHTENED ATTENTION AND FOCUS

Here we can see that Luna is feeling a little insecure in a situation that is unfamiliar to her.

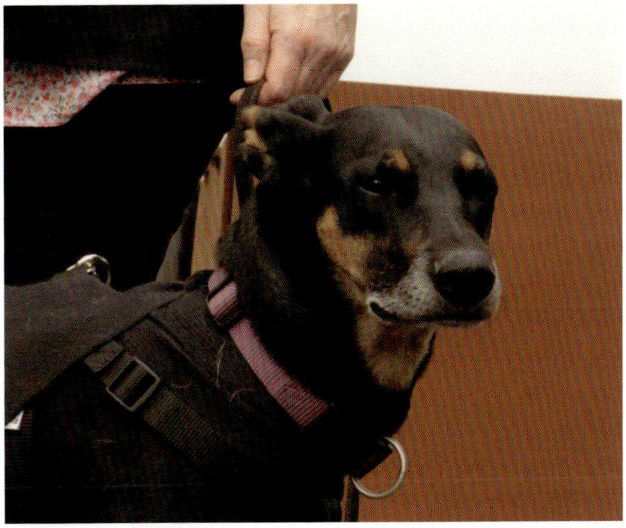

Uneasy and showing discomfort

- Head carriage, and neck, are slightly hunched

Additional clues indicating muscular tension and discomfort:

- Ears pulled back and flat
- Top of the head very tight and smooth
- No direct line of vision
- Mouth closed

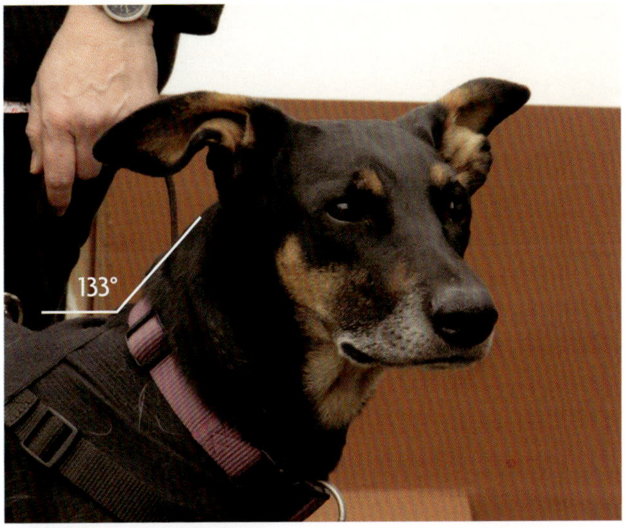

Increased muscular tension and focus

- Neck slightly raised

Additional clues indicating that the muscular tension and focus are increasing:

- Direct line of vision
- Eyebrows slightly pulled towards each other

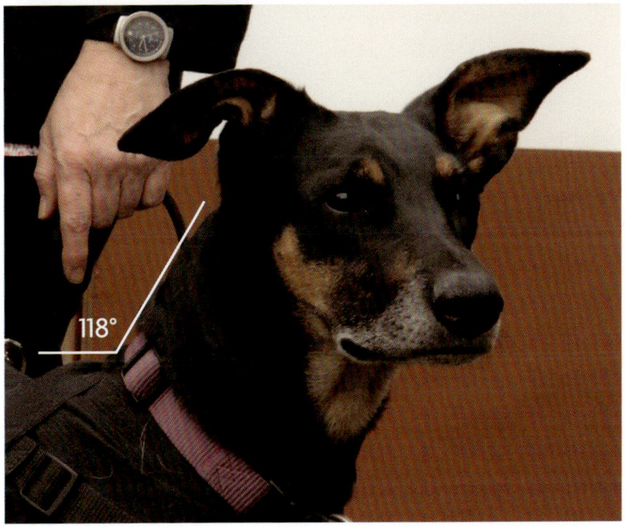

Increased muscular tension with intensified focus

- Neck clearly raised

Additional clues indicating that muscular tension and focus are high:

- Ears are raised at the base and pointed forward
- Eyebrows raised and pulled close together
- Slight wrinkling on the top of the head

This female lives in a rehoming centre and is exercising in an outside pen with a group of other dogs. She is moving without any body tension.

Now she spots the photographer, and raises her head to make eye contact.

CHANGING HEAD CARRIAGE FOR DIFFERENT MOTIVATORS

This male Husky is interested in a female dog, who is ahead of him and slightly to the right. He carries his ears slightly to the side (flirting position) to signal his friendly interest (*see The ears, page 29*).

High head carriage with body tension due to interest

- Neck raised high
- Nose pointing down

Additional clues indicating light tension:

- Whale eye (*see page 69*)
- Mouth closed

Quivive has discovered some rabbits in a snow covered field. She has gone into a hunting mode exhibited by her heightened body tension and raised front paw.

Very high head-carriage with extreme body tension (hunting mode)

- Neck raised high
- Head lifted high

Additional signs of body tension:

- Back rounded
- Ears raised at the base and pointed forward

York, the Golden Retriever, is clearly showing his disinterest with a raised head, in combination with eye contact avoidance, and the slightly raised tail. If the ears are flat and the corners of the mouth are tense, displeasure is involved. Here, York feels a bit annoyed by the ardent expressions of love from Arwen, who has known him from puppyhood. York confidently displays an 'adult' demeanour to gain some respect:

Very high head carriage, indicating minor annoyance

- Neck stretched high
- Head carried high

Additional clues indicating body tension:

- Tail carried level with the back
- Ears flat on the head, and pulled back
- Corners of the mouth pulled back
- No eye contact

YOUNG DOG ADVANCES

The behaviour Arwen (the smaller dog) is showing in this scenario is used to strengthen social bonds between dogs who know each other. However, it may be advisable to call off the youngster if the adult feels crowded and does not know how to diffuse the situation appropriately. Otherwise, this behaviour can quickly turn into a serious predicament.

Be careful if you have an exuberant puppy such as this, who does not understand the importance of giving space to unfamiliar adult dogs! Not all dogs will respond kindly. Pain, chronic illness, a lack of (or negative) experiences with other dogs, or, simply, the belief that such behaviour is rude and lacking respect, can lead to aggression or disciplinary action. We advise guardians of puppies and young dogs to practise polite behaviour when approaching other dogs by respecting personal space.

Rani, the Golden Retriever, wants her housemate, Ella, to settle down. We assume that Rani is either barking or growling. Her point of balance is forward.

Lowered head-carriage while annoyed

Rani (right)

- Head and neck carried below her back

Additional clues indicating physical tension:

- Tail is held level to the back and stiff
- Ears pulled back slightly
- Direct eye contact
- Whiskers or vibrissae raised
- Constricted mouth

This male Rhodesian Ridgeback shows confrontational, even threatening, signals as he approaches another dog. The closer he comes, the more he drops his head. Fortunately, on this occasion, the other dog turned away and returned to his guardian, who had called him several times.

Low head-carriage

- Head carried below the level of the back

Additional clues indicating physical tension:

- Direct eye contact
- Ears pointing forward and raised at the base
- Severe wrinkling on top of the head
- Posture and movement forward
- Hair on top of the neck raised all the way to the tail (even the ridge, typical for the breed, is raised).

This photo, featuring Rani, the Golden Retriever, indicates a certain hesitancy in forward movement, as if she is trying to decide if she should proceed. In this instance, it is highly likely that a wild boar was nearby, although none of the other dogs on the walk noticed it, and did not change their behaviour.

Low head-carriage: attention combined with caution

- Head carried lower than back

Additional clues indicating physical tension:

- Back is straight
- Tail raised
- Point of balance forward
- Line of vision focused
- Ears carried sideways
- Left front leg raised
- Hair raised over the back

Fionn, a Collie mix, is experiencing some rough play from a female Landseer, Fredi. In fact, Fredi is a very nice dog, but can get rough when she plays and, as a result, Fionn shows signs of stress.

Low head-carriage

- Neck is clearly lowered

Additional clues indicating stress:

- Deep wrinkles on the face
- Lowered head
- Ears flat and pulled back
- Top of the head is tightened
- White of the eye is visible (*see page 70*)
- Tip of the tongue is widened

The handler is putting on a search harness, which is used for mantrailing training. The Shepherd's regular handler is injured, so the person handling the dog is a stranger. As he bends over top of the dog to pick up the leash, the dog's body language displays a high level of stress.

Low head carriage when experiencing stress

- Head caried only slightly above topline/line of the neck

Additional clues indicating tension:

- Ears flat and pulled back
- Top of the head tight
- Eyes round
- Eyebrows raised
- Tongue is long and hanging out of the mouth; the midline is visible

HIGH AND LOW HEAD CARRIAGE

The confident, older female (*right*) in this Sloughi breeding facility enters the scene with her head held high. The younger female shows respect towards the older dog by lowering her head. She pulls her ears back and blinks without direct eye contact.

- Both dogs are holding their tails lower than their back
- Both dogs avoid direct eye contact
- The ears of the older dog (*right*) hang loosely
- The young dog (*left*) is blinking and lowering her head

MOVING THE HEAD TO THE SIDE

For anatomical reasons, some breeds are more limited in their ability to move their necks from side to side. This includes short-necked breeds, such as Bull Terriers and Mastiffs, and smaller breeds, such as Pugs and French Bulldogs. For these dogs, it is not as easy to use the communication signal of turning the head away, so they have adapted by turning their whole body. On the flipside, they will also turn their entire body towards another dog, which can come across as more confrontational than just turning the head.

A dog's individual movement and behaviour patterns can cause misunderstandings with another dog which, for example, may have a very flexible neck and would only turn its head towards a strange dog. This is one of the reasons why dogs of the same, or similar, breed usually understand each other better.

Quivive, the Pyrenean Sheepdog, has a very flexible neck.

This female Bull Terrier is clearly less flexible.

The Bull Terrier approaches Quivive and asks her, very nicely, to play. However, with her more basic motor functions, she keeps her body stiff and, instead of turning her head, she turns her whole body towards Quivive.

CHANGING THE PERSPECTIVE

Perspective plays an important role when greeting another dog, and this is particularly the case with smaller dogs. Small dogs, much like small children, see the world from a different angle, which will have an effect on what they see, and how they interpret what they see.

We rarely see our dogs from below. Sometimes when we are playing with them on the floor we may be eye-to-eye, but our perspective is usually from far above them.

In an urban environment, smaller dogs need the protection of their guardian.

Out in the countryside, a small dog's ability to scan the landscape can also be inhibited, usually by tall grass and bushes. This may be the reason why many smaller dog breeds prefer trails and short grass. The notable exception here are small hunting dogs, such as Bassets, Dachshunds and Jack Russells.

5. THE EARS

Hearing is a very complex process which we will simplify for the purposes of this book.

Dogs can hear a greater range of frequencies than the human ear is able to process, so we are not just inferior to our canine friends when it comes to our ability to smell. Humans hear sounds that are between infrasonic and ultrasonic ranges, so between 16 Hz to 20.000 Hz. Dogs, no matter what breed, can hear ultra-frequency sounds of up to a minimum of 50.000 HZ at a distance of about 600 metres (approximately 650 yards or 0.4 miles).

There have not been enough tests on infrasound below 16 Hz in dogs. However the vibrations of high acoustic pressure, created by the infrasonic frequencies, seem to be easily felt. They create physical changes, such as changes in ear pressure, reduced respiration and raised blood pressure. These changes can be symptomatic of fear.

Just as we can become completely absorbed in a fascinating book, and no longer notice our environment, dogs can become so distracted by smells and sounds (which we are not aware of) that they no longer respond when called. Consistent tension and restlessness, without an apparent cause, may be prompted by noises we are unable to hear. Just like humans, dogs respond to stress with an increase of stress hormones and decrease of tolerance.

Individual dogs have different abilities to hear. As the dog ages, his hearing and his eyesight diminish but, usually, his sense of smell remains intact even into a very advanced age.

To summarise, we believe that many of a dog's reactions and behaviours, such as barking, restlessness and fear, may be a response to sounds that we are not aware of, because their frequency is too high for our ears, or they are too far away.

THE OUTER EAR

The pinna (ear flap) and the outer ear canal are part of the outer ear, which is separated into a first and second section. The first section of the outer ear canal goes straight down into the dog's ear and is usually easily visible. We will focus extensively on this part of the ear because it supplies us with a lot of detailed information regarding the dog's emotional state in different situations. The second part of the ear canal is not visible and leads to the eardrum and middle ear.

1. Tip of the ear.
2. Inner edge of the ear.
3. Outer edge of the ear.
4. The pinna (ear flap).
5. Base of the ear inner (top).
6. Base of the ear outer (bottom).

There are lots of different structures that help with the reception of soundwaves and, therefore, improve the dog's quality of hearing. Dogs with drop ears hear almost as well as dogs with ears that are upright. However, dogs with hanging ears have a slight disadvantage when listening for pitch.

When sniffing, dogs with extreme, long hanging ears can hear more sounds that are created at ground level, or even underground.

ONE FORWARD – ONE BACK

The ability to move ears in different directions

There are 17 muscles in the pinna that allow the dog to move his ear like a directional microphone towards the source of a soundwave. He has the ability to move each ear independently, so he can pinpoint the location even more accurately. This also means that a dog can split his attention when his ears are pointed in different directions.

These dogs are splitting their attention between what is happening behind them and the photographer who is in front of them. This is more difficult to see in dogs with long facial hair. Holding the ears in different directions can also be an emotional response, expressing inner conflict, mixed emotions, insecurity or pain (*see Warning facial expression, page 97*).

EAR SHAPES

We will not be able to show you all the different shapes of ears in this book, there are just too many! Here are some of the most common:

NATURAL EAR SHAPES

Erect ears: Pharaoh Hound

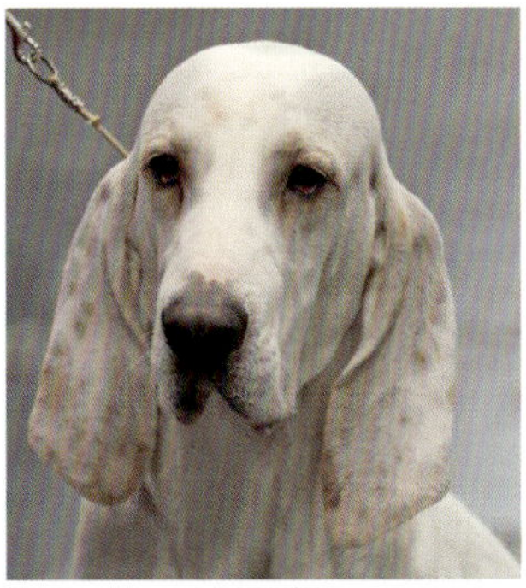

Folded ears: French Porcelaine

V-shaped/dropped ears: Azawakh

Rose ears: Mixed breed

Semi-erect ears: Shetland Sheepdog

Button ears: Pug

UNNATURAL EARS, CROPPED EARS

Sadly in the past, some dog breeds had their ears artificially altered, which means parts of the ear (and/or the tail) were cut off for cosmetic purposes. Ears were cropped any time from day one to seven weeks, because it was assumed puppies of this age did not feel pain. Today we know that experiences and trauma at that age are deeply ingrained in the dog's subconscious memory.

Fortunately cropping ears, and docking tails, is outlawed in many countries. It is now widely recognised that this artificial alteration robs the dog of the ability to express his emotions and to communicate with other dogs.

Spike, the Dobermann (*left*) with cropped ears, and Vera, (*right*) with natural ears. Both dogs are paying attention to the upcoming treat with very different facial expressions which are created by their ears.

There is no doubt that dogs with cropped ears are more difficult for other dogs to read than those with natural ears. It is important to note that there are acupuncture points all along the ear, and especially around the edge, which are connected to the entire body. There is a particular point for calming, located at the tip of the ear that can be stimulated in an emergency. However, the formation of scar tissue along the edge of the ear can cause false sensations (i.e. phantom pain) and eventually impact circulation.

HIDDEN EARS

In some breeds, it is impossible to see the ears due to a very dense coat. This type of coat can cause miscommunication, and it is harder to detect how the dog is feeling. It is, therefore, important to observe body language closely, and also consider the context. If the tail is also hidden by fur then communication with other dogs, even those with experience in reading body language, will be hindered.

You can only guess how these two dogs are carrying their ears. This makes communicating with other dogs very challenging.

EAR POSITIONS IN PUPPIES AND YOUNG DOGS

Four siblings, with three different ear types. In most cases, ear carriage will not be set until puppies have finished teething. This is particularly the case with dogs that go on to have erect ears.

Different moods are easily recognised through the positioning of the ears. Puppy ears are very flexible because the cartilage is still soft. This puppy, Robbie, has remarkable control over his erect ears.

WHAT DO EAR POSITIONS TELL US?

Ears, just like the eyes and nose, are, first and foremost, a sensory organ. Their primary role is to receive acoustic information; their use as a tool for communication is secondary. When communicating, most information is conveyed through the posture, and positioning, of the ear. It is, therefore, important to acquire knowledge of neutral ear positions, and their associated facial expressions.

Generally, the more the ears are pointed, or pricked, the more physically tense (positive or negative) the dog is likely to be. The more the ears are carried down or back, the more insecure or frightened he is likely to be. Of course, there are exceptions, such as when when a dog is extremely excited about something, or extremely enraged (*see Part III: The Emotions*).

RANGE OF EAR POSITIONS

Using erect ears as an example

A dog with erect ears can position them straight up, turn them sideways, or lower them and move them closer to his body. The changing ear position has a major effect on the dog's facial expression, which allows us to pick up on mood. To avoid misinterpretation of the dog's signals, it is important to evaluate all the facial features, rather than the ears alone.

Attentive

- Both ears upright
- The pinna is pointing forward and raised upward at the base of the ear

Conflicted

- Right ear is turned slightly to the side and raised at the base
- Left ear is flatter and pointing to the side

Insecure

- Both ears are turned towards the back and flattened
- Pinna is held sideways and the tips of the ears are pointing down

FROM TENSION TO RELAXATION

Bako, a sable German Shepherd with erect ears, presents very different ear positions depending on his attention level.

Here he displays attentive body tension, paired with vigilance. His ears are pushed closer together because of the tension on top of his head.

The top of the head shows some wrinkling. This expression is exaggerated by the marking on the top of his head and around his eyes. He holds direct eye contact.

Very tense

- Ears are raised at the base
- Pinna is turned towards the front

Additional clues indicating tension:

- Eyes rounded and focused.
- Eyebrows pulled closer together
- Mouth closed

Bako is more relaxed, evidenced by his ears being held farther apart and turned to the side. The top of his head is smooth, and his eye is soft because he is no longer tightening his eyebrows. He is not holding direct eye contact..

Relaxed

- Ears are held to the side at the base
- Pinnae of both ears are turned into different directions

Additional clues indicating relaxation:

- Eyes are almond shaped
- Eyebrows are not pulled together

DROPPED OR V-SHAPED EARS

Short-haired dogs

When dogs have dropped ears, the degree of tension can often only be seen at the base of the ears, and by how close the ears are held to the body. The changes can be very subtle.

This dog's attention is on something ahead of him. His facial expression shows concentration.

Very tense

- Ears are slightly raised at the bottom and held further forward

Additional clues indicating tension:

- Slight wrinkling on top of the head
- Raised eyebrows
- Mouth closed

You can see the relaxation not only in the ear carriage, but also in the smoother skin and loosely hanging upper lip

Relaxed

- Base of the ears not raised
- Ears are loose and fall back a little bit due to gravity
- Ear folds at the edges

Additional clues indicating relaxation:

- Upper lip hangs loosely

FOLDED OR FLOPPY EARS

Long-haired dogs

It is even more difficult to observe the change of position when the floppy eared dog also has long hair. In the following photos, we have to consider all other factors to evaluate the dog's mood.

China, the Cocker Spaniel, ended up in the ocean against her will. You can clearly see by the wrinkles around her nose and eyes, and the closed mouth, that she did not enjoy the experience.

Very tense

- Base of the ears pulled flat back and down.

Additional clues indicating tension:

- Top of the head smooth and pulled down
- Wrinkles on her face
- Lips pressed tightly together

Now China trots along the beach and is relaxed. The base of her ears appear to be raised slightly, which may be due to the wind.

Relaxed

- Ears not raised at the base, nor carried sideways

Additional clues indicating relaxation:

- Mouth slightly open

DIFFERENT EAR CARRIAGE FOR DIFFERENT MOODS

If you are familiar with your dog's natural ear carriage, it is easy to judge how relaxed or tense he is, at any given moment, by observing how the ears are positioned.

FROM ATTENTIVE/RELAXED TO UNCERTAIN

Rose ears

Lotti is a female mixed breed, with partially raised ears. The tips of her ears fall to the side and back. When looking from the side, the shape of her ear is similar to rose ears. Here she is clearly relaxed, and is looking with interest and attention, at Katja who is positioned in front of her.

Relaxed and attentive

- Ears are raised sideways at the base and turned slightly towards the front
- Tips of the ears point down

Additional clues indicating she is attentive and relaxed:

- Eyes slightly almond shaped
- No tension in the eyebrows
- No wrinkles on top of the head
- Midline above the eyes clearly visible

Lotti is in the presence of unfamiliar people – she is always extremely tense when meeting strangers.

Extremely tense

- Ears pulled back at the base and pasted flat against the head
- Ears are almost invisible when looking at her from the front
- Pinnae are turned extremely far to the back

Additional clues indicating extreme tension:

- White of the eye is visible; the eye is small
- Puffed cheeks
- Top of the head is tight and pulled back
- Midline above the eyes is almost invisible

FROM ATTENTIVE/RELAXED TO VERY TENSE

Semi-erect ears

Fanny, a female Berger des Pyrénées, had very little experience with different environmental stimuli before arriving at Gabi's at the age of three. In unfamiliar situations she was often unable to cope. Here she looks attentively at Gabi.

Relaxed and attentive

- Ears are partially raised at the base
- Tips of the ears hang loosely to the side

Additional clues indicating relaxation:

- Direct eye contact
- Slightly almond-shaped eye
- Erect body posture
- Tail is moving

Fanny (*left*) appears to have no ears. This primarily occurs in situations when someone gets too close, or it could be in relation to the camera. In contrast, her companion Tiber (*right*) shows relaxed, loose ear carriage, with raised ears. Both dogs are slightly leaning away from each other. (*See Switching Balance point, page 162*).

Tense

- Base of the ear pinned back
- Base of the ear not visible

Additional clues indicating tension:

- Mouth closed
- Slight crouching

FROM RELAXED TO TENSE

Erect ears

This female mixed breed lives in a shelter. She is just getting her harness adjusted to go for a walk. The handler respects the space the dog needs while she is dealing with the adjustments of the harness. The dog's face is without any tension, she is very comfortable with her handler.

Relaxed

- Ears raised at the base, turned slightly to the side
- Pinnae turned softly forward

Additional clues indicating relaxation:

- Mouth slightly open
- Eyes almond shaped

Now someone has grabbed and pulled her collar, and the position of her ears reveals her tense, emotional state. In addition, her whiskers (vibrissae) around her muzzle are pointing slightly more forward. It is clear that this situation is making her feel uneasy.

Tense

- Ears pulled back at the base and flattened
- Pinnae flattened and turned towards the back, more on the right than the left

Additional clues indicating tension:

- Eyes small
- Tip of the tongue visible

FROM WIDE AWAKE TO TIRED

Tamika is looking in the direction of the camera and is interested in what is going on.

- Ears are raised at the base
- Ears are turned forward

Now Tamika is starting to get tired. Her ears, which are usually erect, are starting to relax to the side of her head. She carries each ear differently; the right ear to the side, the left ear is flat and pointing towards the back.

- Right pinna turned forward
- Left pinna turned towards the back and down

This photo was taken just moments before Tamika closed her eyes.

- Left ear held upright and sideways
- Pinna turned forward
- Right ear still slightly raised at the base
- Two-thirds of the ear folded forward

FROM CONFRONTATIONAL TO FRIENDLY/INTERESTED

Ben, an entire mixed breed male, approaches Quivive (out of shot) in a confrontational manner. As he gets closer he raises his head, his tail is stiff and points straight up. He doesn't know Quivive and, therefore, seeks to intimidate her.

Confrontational

- Ears raised high at the base
- Ears pointed forward

Additional clues indicating tension:

- Tail carried high
- Head dropped and tensed
- Direct eye contact
- Mouth closed
- Dynamic forward movement

Ben has realised that Quivive is a female and changes his ear position immediately. He raises his ears high at the base and turns the pinna back in a typical flirting position.

Friendly/interested

- Ears raised high at the base
- Pinna turned to the side and back

Additional clues indicating friendly interest:

- Head is carried lower and softened

Find out more about this meeting between Ben and Quivive, see page 62.

FROM TENSE – TO VERY TENSE – TO RELAXED

Luke, the adult Border Collie (*right*) is meeting a new addition to the family, a four-month-old pup named Jake. Luke is insecure and does not know how to respond to Jake, who is very pushy.

Luke is annoyed by Jake's impolite intrusiveness. He looks past the youngster avoiding eye contact, and is clearly very uncomfortable with the puppy invading his personal space.

Tense

- Ears flat at the base and pointed to the side and back
- Pinnae turned to the back
- Tips of the ears held horizontally

Additional clues indicating tension:

- Slit-shaped eyes
- Eyebrows pulled towards each other
- Mouth closed

Luke now gives a warning by showing his teeth. He turns his eyes towards Jake; his opened mouth is turned away from the puppy.

Very tense

- Ears pulled back at the base and flattened
- Pinnae turned all the way back

Additional clues indicating extreme tension:

- Slit-shaped eyes
- White of the eye visible
- Eyebrows pulled towards each other
- Top of the head tightened
- Teeth exposed

Following some interactions with Jake, which were more harmonious, Luke is feeling less stressed and agitated.

Relaxed

- Ears flat at the base and carried sideways
- Pinna turned to the side.

Additional clues indicating relaxation:

- Eyes less slit shaped, rounder
- No tension in his upper lip
- Tongue hangs loosely from his mouth

A TIME TO FLIRT

When dogs are interacting with each other, they may show flirtatious behaviour. This is what it looks like, with particular reference to ear position:

- Ears slightly raised at the base and carried towards the back
- Pinna turned to the side
- Tips of the ears pointed downward

In this photo, the female Golden Retriever, Rani, meets a tan coloured male for the first time and she likes him. Initially, the male is reserved and then wants to make contact with Rani. He displays the typical ear position of a male who is very interested in a female, which is known as the 'flirting position'.

Here, the tan coloured male shows typical flirting behaviour towards Tamika. (*left*).

Arwen (*left*) shows that a female also displays flirting signals when she is interested in a male.

Arwen (*right*) meets a male who wants to connect with her. You can see that he is polite and friendly by the flirting position of his ears. He also shifts his point of balance away from Arwen.

It is harder to spot the flirting position in long-haired dogs, but it is still possible if you know what else to look for.

Male (*left*):
- Ears are raised at the base and held sideways
- Pinna is turned to the side and back

ON THE MOVE

Strange things can happen when dogs are interacting on the move!

6. THE EYES

A dog's moods and emotions can also be communicated by the size and shape of his eyes, his pupils and the positioning of his eyebrows. To interpret what the eye is trying to tell us, it is important to be familiar with the neutral position of the eye when the dog is relaxed, happy and feels safe.

Dogs have round or slightly almond-shaped eyes, although some breeds, such as the Japanese Akita, have slit-shaped eyes. If the eyes appear larger than normal, it can be a sign that the dog is excited or very impressed by something he is experiencing or seeing. He may also be feeling fearful or stressed. An aggressive dog often displays a larger eye than normal (for him), or when really angry, he can reduce the size of his eyes to a slit.

In poor light conditions, the dog's vision is far superior to that of a human, because the eye behind the retina has a special light reflecting layer, the tapetum lucidum, that humans lack. This layer also causes the dog's eye to glow in the dark when it is hit by light. The colour can vary.

Dogs perceive colours differently from us. They primarily see in the blue-yellow range, which is similar to people who suffer from red-green colour blindness. The position of the eyes, on the side of the head, allow for superior peripheral vision, although this varies from breed to breed. We humans are able to see objects more clearly at a distance. Movement makes a big difference, and dogs can spot moving objects more clearly than those that are stationary.

It is important to be aware of all these factors when we are attempting to pinpoint what might be unusual behaviour, as well as interpreting the emotional response.

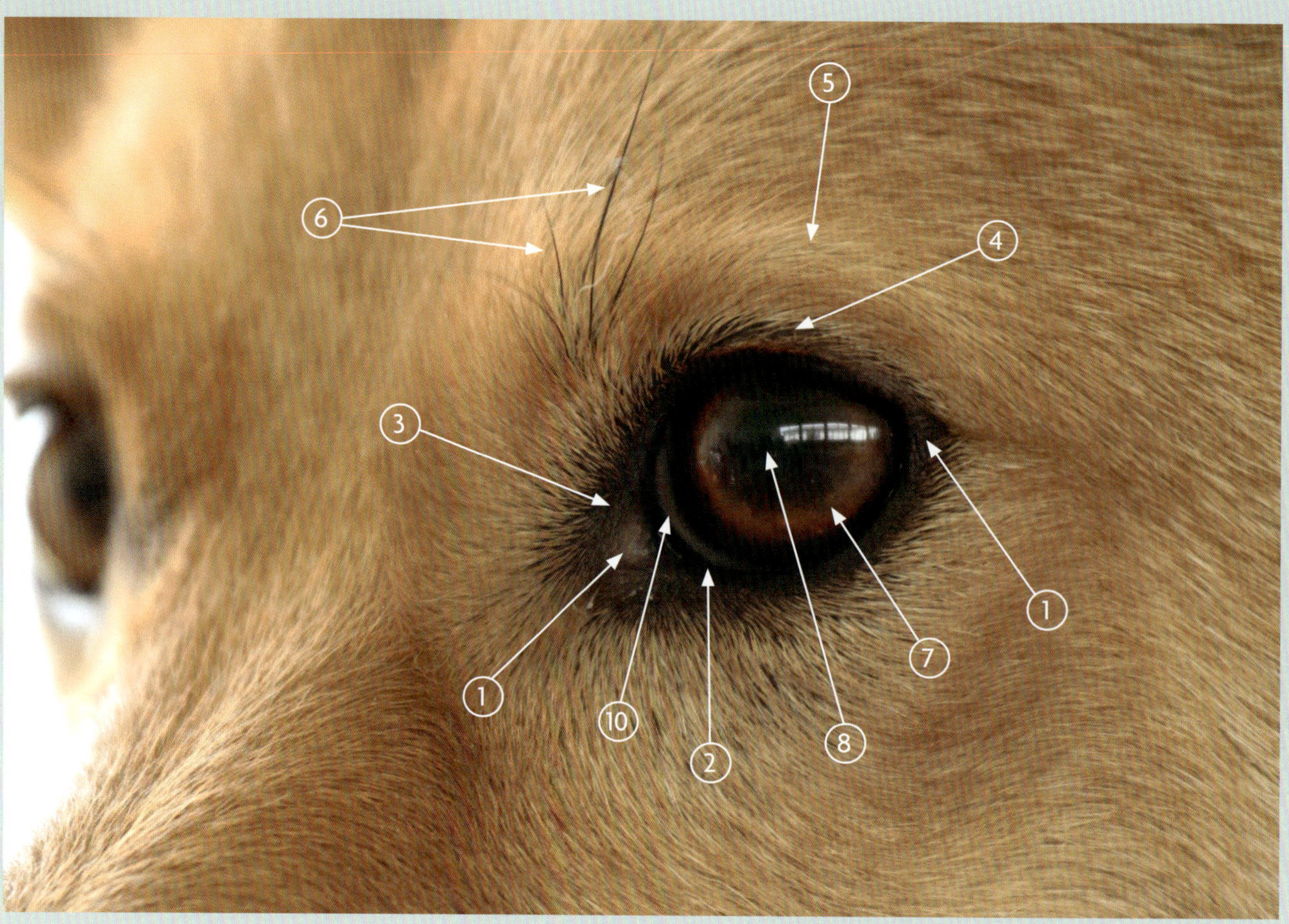

Visible details of the eye

1. Inner and outer corners of the eye (they determine the shape of the eye).
2. Eyelid.
3. Hairless area.
4. Eyelashes *(also see page 47)*.
5. Eyebrow.
6. Whiskers (vibrissae or sinus hairs).
7. Iris.
8. Pupil.
9. The white of the eye (corium, sclera): *see page 47*.
10. Nictitating membrane (third eyelid).
11. Conjunctive tissue *(see page 47)* – quite inflamed in this dog.
12. Raised nictitating membrane (third eyelid): *see page 47*.

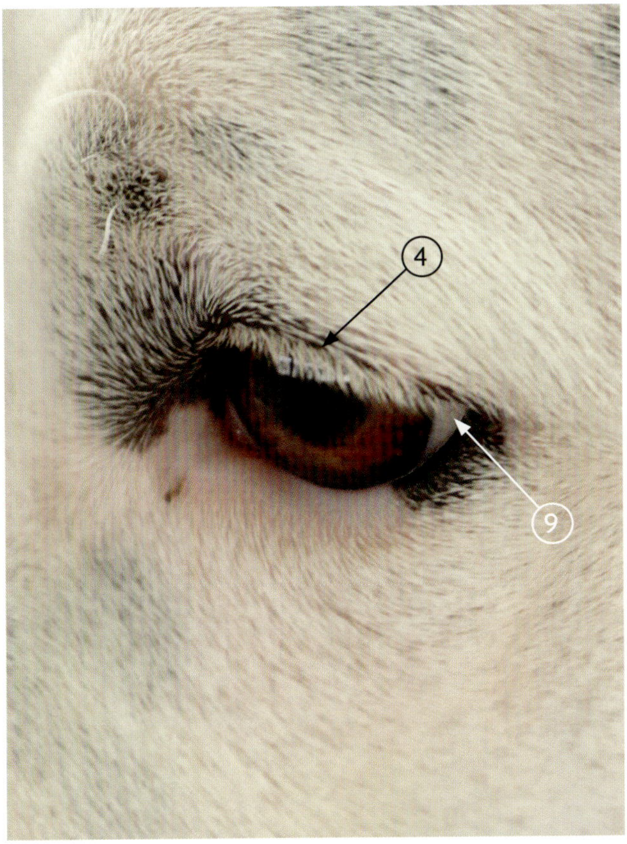
4
9

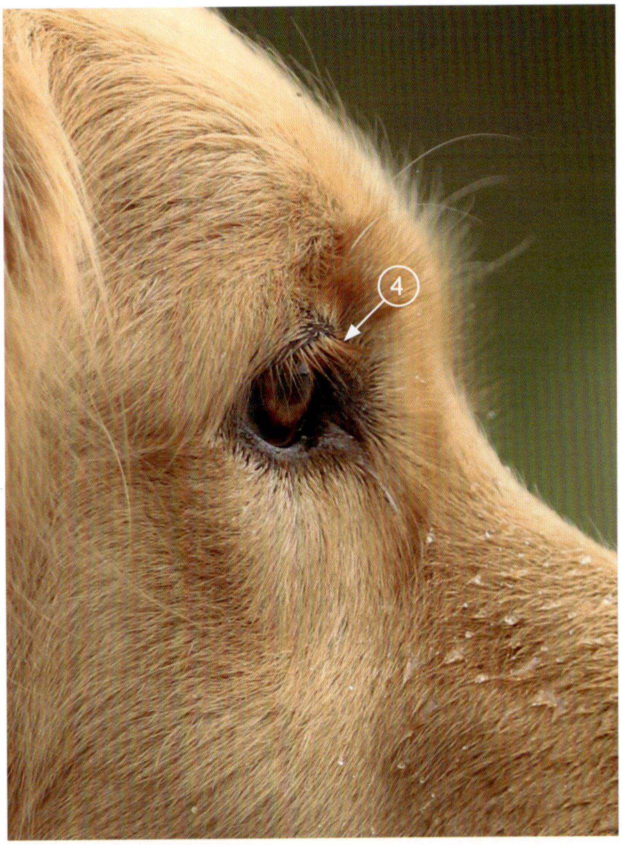
4

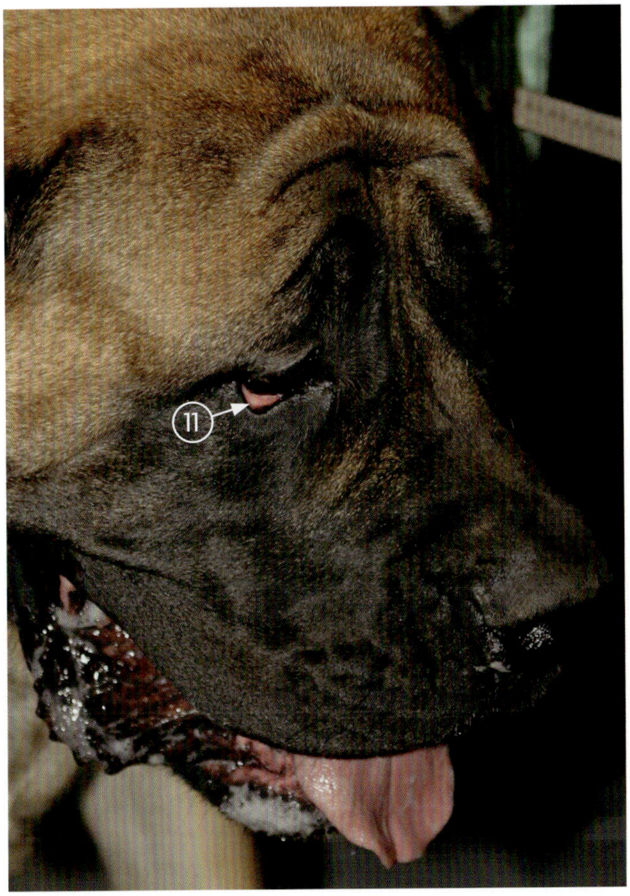
11

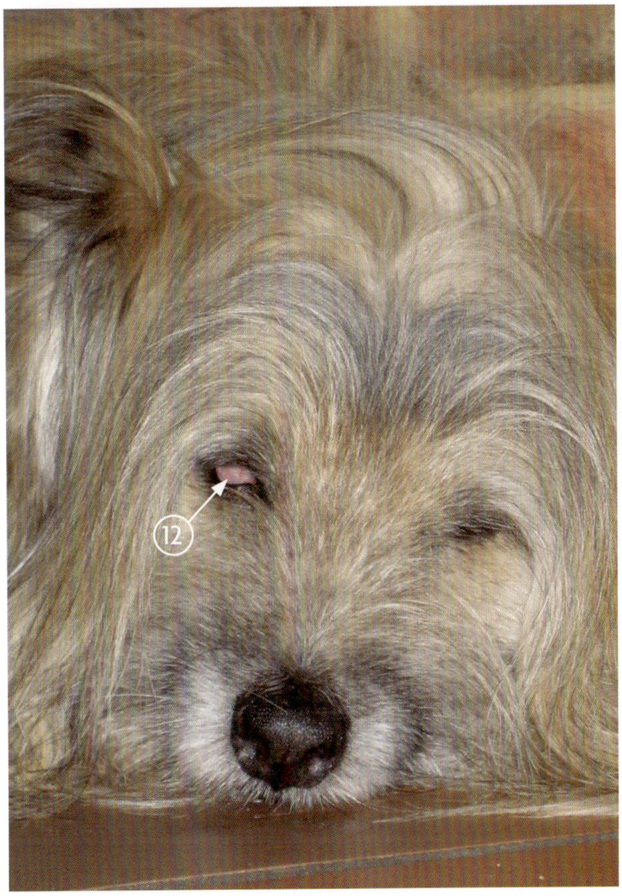
12

EYE SHAPES

Almond-shaped eyes.

Slightly slit-shaped eyes in mother and puppy

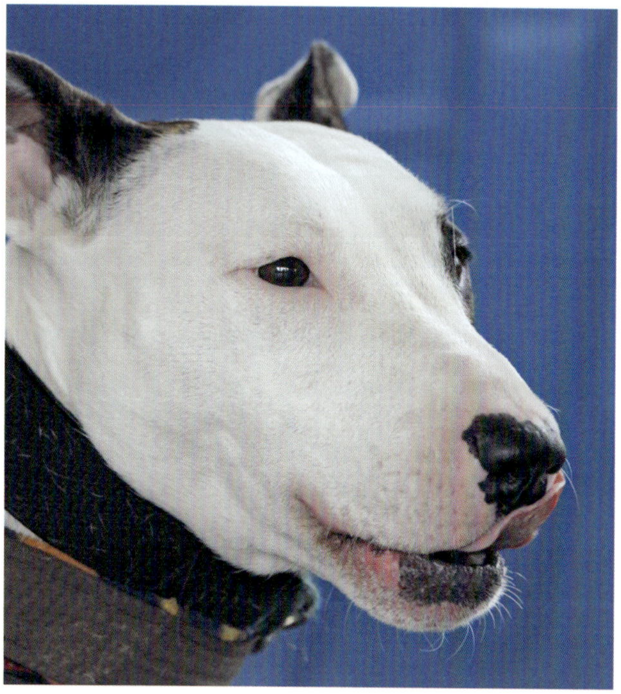

Slit-shaped eyes

Rounded eyes

FROM TENSION TO RELAXATION

In dogs with differently-shaped eyes

Dogs with differently-shaped eyes will often have quite different expressions, even when they are displaying the same emotions.

ALMOND-SHAPED EYES

Quivive, the Pyrenean Sheepdog, is pictured on a walk when she is feeling relaxed. This type of expression is often described as a 'soft eye'.

Relaxed

- Almond-shaped eye
- Not looking at anything specific

Additional clues indicating relaxation:

- Ears slightly raised at the base and held sideways
- Tips of the ears fall loosely
- Mouth closed
- No tension in her lips

Quivive has encountered a potbellied pig for the first time and is feeling very insecure.

Tense

- Very rounded eye
- Eyebrows raised

Additional clues indicating tension:

- Ears raised at the base of the ear and pointed towards the back
- Tips of the ears tightened and held back
- Skin around the nose tightened with hairs slightly raised
- Mouth closed
- Upper lip does not cover the lower lip
- Nostrils widened

SLIT EYES

Quivive (out of shot) is approaching this female Bull Terrier who is looking on with interest. There is no tension in the skin around her mouth and she would like to make contact.

Attentive – slightly tense

- Lightly rounded eyes in the area of the upper lid
- Eyebrow slightly raised at the inner corner of the eye

Additional signs indicating slight tension:

- Ears pointing forward
- Slight wrinkling on top of the head between the ears

Now the Bull Terrier is tense but the changes in eye shape, and positioning of the eyebrows, are difficult to see due to her low set eyeballs and white coat colour. We, therefore, need to observe the position of her ears, the carriage of her tail, and the shape of her mouth, in order to gain a more accurate picture with which to interpret her emotions.

Insecure – tense

- Eyes a little smaller

Additional clues indicating tension:

- Ears held flat pointing towards the back
- Top of the head tight and no wrinkles
- Mouth lengthened
- Whiskers slightly raised

ROUND EYES

In dogs, a more rounded eye signals excitement and tension, but if a dog has naturally round eyes, it is hard to detect these feelings. We may observe that the white of the eye is more visible, or the eyeball is slightly more protuberant. However, in cases where the eyeball clearly protrudes from the eye socket, the dog is experiencing extreme stress, and needs immediate help and support.

Attentive – relaxed

- Eyeball is deep in the eye socket
- Eyebrows only slightly raised

Additional signs indicating relaxation:

- Ears carried to the side
- Top of the head flat
- No excessive wrinkling in the face
- Upper lip hangs loosely
- Mouth closed

Tense

- Eye more rounded
- Raised eyebrow is noticeable
- White of the eye is visible

Additional clues indicating tension:

- Wrinkles developing between the eyes
- Tongue pushed out, held tightly
- Ears turned towards the back

ACTIVE RELAXATION THROUGH TELLINGTON TTOUCH®

There are some dogs that live with a high degree of tension and often react very quickly, and impulsively, when their situation changes. This is often the case with dogs which hard focus on other dogs, or those that feel they are the targets of hard focus. The Tellington TTouch® method offers techniques to actively relax the dog. These relaxations can also be observed in the positioning of the eyebrows and the size of the eyes (*see The Body, page 190*).

Many dogs become tense when they are touched on their legs and feet, but this may be necessary – for example, during a veterinary examination. Here, Roxi is showing signs of tension when, initially, handled by a Tellington TTouch® practitioner.

Tense

- Whale eye (white of the eye visible) due to direct eye contact with the practitioner who is located above.
- Eyes small
- Left eyebrow raised
- Head is not resting comfortably

Calmly and mindfully, the practitioner continues to TTouch® along Roxi's leg and we can now clearly see he is feeling relaxed.

Relaxed

- White of the eye no longer visible
- Eyes more almond shaped
- Left eyebrow lowered
- No longer making direct eye contact
- Head is resting relaxed on the ground (*see page 74*)

POSITIONING OF THE EYEBROWS

The positioning of the eyebrows can tell us a lot about a dog's mood. To interpret how he is feeling, we need to be able differentiate between a tensed and a relaxed eyebrow. Similar to humans, dogs can pull their eyebrows together in anger, or raise them when surprised or curious. When a dog is fearful, his eyebrows are usually raised extremely high, and he will have a very rounded eye.

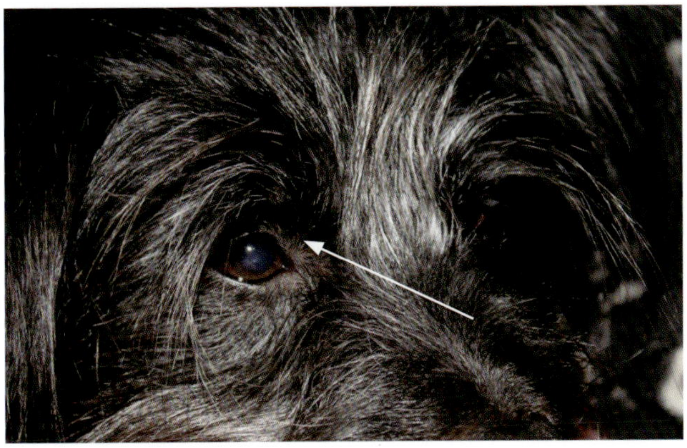

Quivive, Pyrenean Sheepdog

The hairs are shorter, and appear darker, towards the inner corner of her eye (above the hairless area) as indicated by the arrow. Her hairs are longer from the middle towards the outer corner of her eye and fall sideways over the eye. Therefore the eyebrow is noticeable, despite the long facial hair.

Nash, Nova Scotia Duck Tolling Retriever

Even though Nash's coat and skin-coloured pigmentation are the same colour, his eyebrows are defined by the hair showing a change of direction. Note the inner corner of the eye, where the hair changes direction, and creates a line of darker hair above the eye. This makes it easy to see the position of the eyebrows.

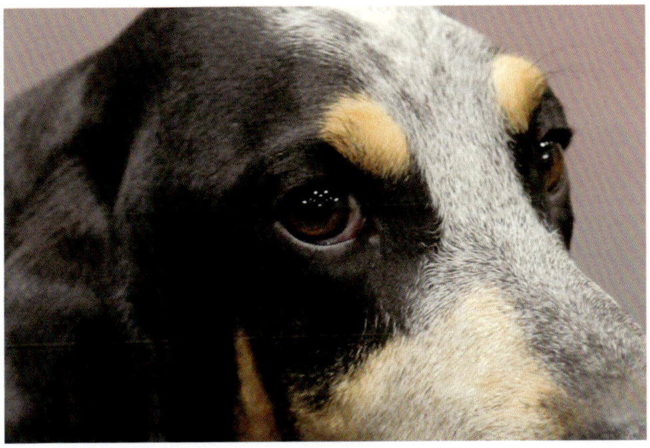

Basset Bleu de Gascogne

Eye spots that are part of the coat markings (*see page 74*) emphasise the position of the eyebrows and, thus, the facial expression. Communication with other dogs depends on how the eyebrows are defined, and what colour they are. In some cases, the eyebrows signal more strongly than the dog intends, and this can lead to miscommunication.

READING EYEBROWS

Eyebrows will be positioned differently, depending on a dog's emotional response to a situation.

RELAXED EYEBROWS

Long facial hair

Quivive is showing a relaxed 'soft eye' which is marked by an observant expression and eyebrows in a neutral position.

Relaxed

- Almond-shaped eye

Additional clues indicating relaxation:

- Ears held sideways at the base
- Tips of the ears hang down loosely
- Mouth closed

TENSED EYEBROWS INDICATING ANNOYANCE

Here Quivive is showing her annoyance as the young, female Eurasian constantly pesters her to play.

Tense

- Eyes rounded
- Eyebrows raised at the inner corner of the eye
- Direct eye contact towards the dog that is passing

Additional clues indicating tension:

- Ears raised at the base, pointing towards the back
- Mouth closed
- Head lowered

RELAXED EYEBROWS

Short facial hair

Spot, Australian Shepherd-Collie
An attentive glance with eyebrows in a neutral position ('soft eye').

Relaxed

- Almond-shaped eye
- No wrinkles between the eyes

Additional clues indicating relaxation:

- Ears hang loosely, left ear slightly back
- Top of the head flat, no tension
- Lips hang down loosely
- Lower teeth visible
- Tongue rests loosely on the lower teeth

TENSED EYEBROWS

Short facial hair

A wild game ensues between Rani (*behind*) and the young German Shepherd in the foreground. We see the 'worried' eye of the Shepherd whose expression is exaggerated due to his dark eyebrows.

Tense

- Eyebrows pulled together and raised
- Rounded eye
- White of the eye visible

Additional clues indicating tension:

- Ears turned towards the back
- Slight wrinkling on the top of the head
- Corners of the mouth pulled back

FROM TENSE TO NEUTRAL

Dhanyi, the female Golden Retriever (*left*), would like to take possession of Arwen's stick. She pulls the stick towards her which causes her neck to tighten and raise.

Tense

- Eyebrows pulled together and raised on the inside corners of the eye
- Direct eye contact
- White of the eye visible

Additional clues indicating tension:

- Ears pulled back
- Top of the head tight
- Neck raised

Dhanyi is holding the stick but no longer pulling on it. Her eyes are less round and her facial expression is more relaxed.

Neutral

- Eyebrows lowered, nearly neutral
- Eye contact nearly broken
- White of the eye barely visible
- Eye less round

Additional clues indicating relaxation:

- Ears hang loosely
- Top of the head loose, midline visible
- Neck lowered

FIXATING

Fixating, which involves intense, and sustained eye contact, often happens very subtly so it may be that the subject is not, initially, aware of it. It can be displayed from a great distance or at close range. There are often no additional changes in the facial expression. Sometimes the eyebrows are pulled a little closer together, the ear position may change slightly, and there may be evidence of tension (the first signs of wrinkling) around the mouth and nose. You can usually observe a stillness; the fixating dog no longer moves, his eye level does not change, and he no longer blinks. This stance is considered a subtle warning. It is important to be aware of the gravity of this behaviour and to respond accordingly. The dog is serious about his warning; he may feel crowded, be guarding a resource – food, a toy, or a person – or he may be preparing to defend a friend.

FIXATING IN THE EARLY STAGES

Both dogs are reacting to a person approaching them. The dog (*left*) freezes whereas the dog (*right*) may be at the point of launching an attack.

- Direct eye contact with the eyes pointing upwards
- Ears held to the side at the base.

OBVIOUS FIXATING

Kaylee (*left*) tells the dog, who is standing in front of her, not to come any closer. She is doing this without moving. This is usually the case with subtle, but clear, fixating or warning.

- Direct eye contact
- Ears pointed forward
- Mouth closed
- Head dropped

Taschka (*right*) warns Tamika not to come any closer to 'her' rat.

- Close to direct eye contact
- Ears pointing forward
- Mouth closed
- Head lowered

Arwen, a youngster (*right*) has got too close to Taschka. Taschka's expression is clear: Stay away!

- Eye rounded
- Right eyebrow pulled upwards on the inside
- Direct eye contact

Additional clues of fixating:

- Ears held flat and to the back
- Mouth slightly open
- Head slightly raised

UNCONSCIOUS FIXATING

Not every dog who makes eye contact is aware that he is fixating. Some dogs are simply curious, interested in connecting, or concentrating on a task or an object. If this sustained focus is directed at another dog, who is unable to get away (for example, in the vet's waiting room), a relatively harmless look can be interpreted as provoking or threatening, depending on the other dog's experience and spatial limitations.

In this situation, the best strategy is to break the dog's focus, by calling his name or gently touching him.

Here, Eve is showing very direct eye contact which has nothing to do with fixating. She is concentrating on her own, very specific, goal.

HARD EYE

A threatening, frozen eye is known as a 'hard eye'. This describes a very tensed eye, seen in an emotionally precarious situation. A dog who shows a hard eye is very serious with regard to his message. He will apply direct eye contact, but also look right through someone, or look past the problem (person or dog).

The difference between a hard eye and one that is widened in fear can be seen at the edges of the eyelids. In most cases, a dog with a hard eye does not fully open his eye. He does not blink, and there is no change in his mouth. His respiration flattens as if he is holding his breath. He may hold his head lower, and will not move his body. There is a stillness signalling calm before the storm.

When a dog directs this look at a person, it should be taken as a serious warning. The person involved should not come any closer or stop whatever they are doing to the dog. In this situation, the dog is no longer requesting, he is threatening.

In fact most dogs who fixate will not actually attack. However, if the dog has had enough negative experiences, and has bitten in the past, he will probably bite again when he shows a hard eye.

If children are present or interacting with the dog, it is very important that supervising adults recognise the developing situation and intervene immediately to prevent an attack.

To recognise a 'hard eye', it is useful to know what a 'soft eye' looks like.

SOFT EYE

In this photo, Daisy, a black, mixed breed female, is calm and relaxed.

Relaxed face with a soft eye

- Slightly almond-shaped eye
- Position of eyebrows is neutral

Additional clues indicating relaxation:
- Ears barely raised at the base and hang down loosely
- Mouth slightly open

Daisy (the black, mixed breed female) is a friendly, peace-loving dog and always tries to settle disputes between dogs. Here she tries to divert Maiky (the Vizsla in the foreground) who has been pestering Benny, who is lying down. Daisy's eyes are very tense and she seems to be staring. However, her intention is not to attack. She is dividing the two dogs by physically moving between them. Her body is directed towards Maiky (the Vizsla) who is the instigator. Daisy is serious in her warning and Maiky, understanding the message, turns away from Benny. He responds by turning his head away so he does not provoke the situation further.

Tense face with hard eye

- Large rounded eye
- Raised eyebrow
- Whale eye (white of the eye visible)
- No direct eye contact

Additional clues indicating tension:

- Slight wrinkles around the nose
- Head lowered
- Ears pulled back

IDENTIFYING THE SIGNS

It is important to recognise, in the early stages, when a dog is showing a hard eye. In most cases, it is combined with the body remaining still. A brief, emotionally charged, situation can often only be detected by a change in the shape of the eyes, and movement of the head.

SOFT EYE

Quivive's eyes are very relaxed. When looking at the following photos, note the changing shape of her eyes.

Relaxed, soft eye

- Almond-shaped eye.
- Neutral eyebrow position.

In this photo, Quivive is keeping a pushy male dog, Ben, at a distance (*see page 45*). She bared her teeth to make her intentions clear, and then looked away to show her lack of interest.

Initiating a hard eye

- Eye rounded
- Eyebrow raised at the inner eyelid

Additional clues indicating tension:

- Ears held towards the back
- Mouth almost closed
- Upper lip does not cover the lower lip
- Head turned away

Since looking away is not working, Quivive turns her head towards Ben without making direct eye contact. She stays motionless, does not blink, and stares.

Hard eye without direct eye contact

- Eye rounded
- Eyebrow raised at the inner eyelid
- Size of the eye unchanged

Additional clues indicating tension:

- Ears pulled back
- Head slightly raised

Quivive now makes direct eye contact, which underlines her demand for space. She stays motionless, doesn't blink and her stare remains hard and fixed.

Hard eye with direct eye contact

- Eye rounded
- Eyebrow raised at the inner corner of the eye
- Size of the eye unchanged

Additional clues indicating tension:

- Ears pulled back
- Head raised

Ben has stopped approaching so Quivive has broken her direct eye contact by turning her head. Her next signal would be to blink in order to de-escalate the situation.

Start of soft eye

- Eye less rounded
- Eyebrow raised inner corner of the eye

Additional clues indicating the start of relaxation:

- Ears pulled back
- Head lowered slightly

Ben has turned away and Quivive's glance softens noticeably. She can now blink because she has released the tension in her lower eyelids. Her movements are still minimal.

Nearly soft eye

- Eye gets smaller, almost almond-shaped again
- Eyebrows raised only slightly

Additional clues indicating light tension:

- Ears slightly forward
- Head carriage unchanged

SUBTLE WARNING

Identified by eye contact and positioning of eyebrows

This photo of the older Malinois female, Nele (*left*), and the youngster, Yaska (*right*), shows their differing emotional states. We cannot see the toy, alongside Nele, which has piqued Yaska's interest.

The positioning of Nele's eyebrows tell us that her "stay away" warning is serious . The two longitudinal folds, forming around her nose, confirm the message. Yaska understands. The position of Yaska's eyebrows could be interpreted as worried. She would really like the toy, but does not want to provoke the older female. Nele's warning is subtle; it is also a typical behaviour that dogs offer towards humans. *For more information on subtle warnings, see The Mouth, page 78.*

Nele (*left*):

- Eyebrows raised and pulled together
- Eyes rounded
- White of the eye visible
- Direct eye contact

Additional clues indicating tension:

- Ears pointed forward
- Wrinkles on top of the head
- Mouth closed
- Wrinkles on the side of the nose
- Head raised slightly

Yaska (*right*):

- Eyebrows raised
- Eyes rounded
- White of the eye visible
- Looking away (slightly)

Additional clues indicating tension:

- Ears turned to the side
- Top of the head flat
- Mouth closed
- Neck raised
- Head lowered slightly

INTERPRETING THE WARNING

Understandably, children often fail to interpret the subtle "stay away"warning, or the threat that is signalled by a hard eye. However, sadly, the same applies to many parents and even experienced dog handlers. The transition from a threatening hard, fixed stare to a bite is fluid and depends on a lot of factors, including:

- The dog's previous experience with children or adults
- Whether he is protecting personal space
- Whether he is on or off leash
- Breed characteristics
- Individual temperament
- The response of an older dog that needs to discipline a pushy puppy (or child)
- The need to defend a resource which can be food, a toy, a person or another dog.

Dogs who are in pain, or those that have difficulty in getting up (which means it is hard to leave a situation) rely on subtle warnings. It is helpful to give these dogs a safe zone where they can rest undisturbed. Many guardians mistakenly believe that a dog should not be allowed to growl at a person, or another dog, and should be punished. However, if the growl follows a subtle warning, through eye contact, it should be interpreted as a means of communication. There is still enough time to respond appropriately to de-escalate the situation, i.e., by removing the dog from the situation and allowing him a chance to breathe. It is also important for the dog's guardian to pause for a moment and think about what they can do to change the situation in order to help the dog.

However, in some situations, it may be necessary to react quickly, and create a distance between the dog and the adult, or child, he is threatening, to avoid an attack.

Problems arise if a dog has been punished for growling in the past. If he is inhibited from using this subtle warning, and is forced to escalate his behaviour to a bite, the situation can no longer be solved positively. We would always recommend using positive training methods (such as clicker training) and working with an experienced positive dog trainer or Tellington TTouch® practitioner.

FROM A POLITE REQUEST TO A SERIOUS WARNING

Polite request for distance: Air-licking, blinking, pawing, yawning, head avoidance.
Urgent request for distance: Ears back, lying down, walking away
Rising stress: Crouching, tucked tail, freezing.
First stages of warning: Freezing, stillness, fixating with a hard eye, closed mouth.
Clear threat to escalation: Growling, showing teeth, snapping, biting

This is just a guideline; the transitions are, of course, very fluid.

IRIS AND PUPIL

The pupil is a colour-infused muscle that can change its size. It is usually mid to dark brown and, in dogs, the outside edge of the pupil is darker.

FUNCTION

The iris has the ability to contract inward when there is too much light, which makes the pupil in the centre of the eye smaller, limiting the amount of light that enters the eye. When this happens, more of the coloured iris is visible. In low or dim light the pupil stretches all the way to the edge, which makes it appear larger, and it allows more light to enter the retina.

The signals to narrow or widen the pupil are sent to the visual centre of the brain through the optic nerve in the iris. This reaction protects the light sensitive retina inside the eye from high intensity light exposure. It operates in the same way as a lens on a camera

Needless to say, the eyes' most important function is sight, but the dog's emotional state can also be detected by the size of the pupil.

EFFECT OF DIFFERENT LIGHT CONDITIONS

The size of the pupils change depending on the light conditions. When there is a lot of light the pupils are smaller, in low light they are larger.

Small pupils due to sunlight.

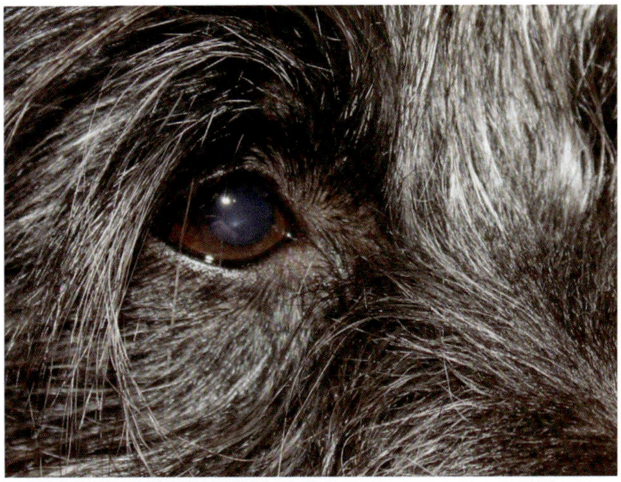

Large pupils in low light.

This German Wirehaired Pointer is clearly excited while attending a show and is searching for more contact with his guardian by standing up and placing his paws on their arms.

Enlarged pupils due to stress or excitement

- Enlarged/dilated pupils
- Not much of the brown iris is visible

Additional clues indicating tension:

- Ears pulled back
- Top of the head is tight
- Tongue widened and very long, hanging from the mouth
- Deep ridge in the middle of the tongue visible

Dawina, the Dalmatian, is stressed during a treatment on her back. Her eye is very round and she looks at the person sitting in front of her with a dilated pupil. The area around the outside of her eye is tensed.

Tension

- Enlarged/dilated pupil
- Rounded eye
- White of the eye is visible

Additional clues indicating tension:

- Ears pulled to the back
- Top of the head is tight
- Upper lip is tight.
- Corner of the mouth is pulled down

If a dog is highly excited, the pupil will widen (dilate) and appear larger, regardless of the light conditions. In this situation the widening is influenced by the sympathetic branch of the vegetative nervous system. In addition to other stress signals, the size of the pupil is a valuable tool to evaluate emotions and stress levels.

IDENTIFYING A WIDENED PUPIL

Many dogs have a brown border around the pupil which can be misread as a dilated pupil.

This English Springer Spaniel is directing his gaze at his guardian. Looking at the shape of his eyes, and noting the whites of his eyes and the pupil, which appear to be widened due to the brown edge, we could easily assume he is stressed. However, the loosely hanging skin on his head tells a different story.

Relaxed

- Pupils seem large
- Eyes rounded, nearly lemon-shaped
- White of the eye visible

Additional clues indicating relaxation:

- Ears hanging loosely to the side
- Upper and lower lips hanging loosely
- Tongue relaxed in the mouth
- Mouth slightly open
- Hardly any wrinkles around the corners of the mouth.

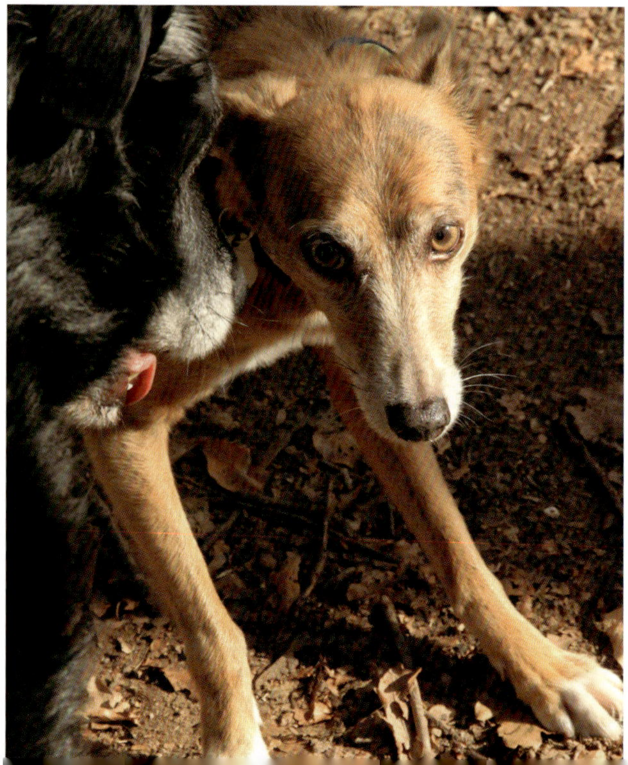

Chita, a Spanish street dog, has a dark border around her pupil, which appears to be narrow in this photo, most likely because of the sunlight that is hitting her eye. She is scared and excited about meeting other dogs so you would expect to see a large, dilated pupil.

Tension

- Moderately round eye

Additional clues indicating tension:

- Ears pulled back
- Crouching

WHALE EYE

When the white of the eye is visible, it is known as a whale eye. Unlike the human eye, the white of the eye is, generally, not seen in dogs because the iris covers the entire surface of the eye that is exposed. Some dogs, with particular eye shapes, show the whites of their eyes when they peek around a corner, or open their eyes extremely wide.

When reading communication signals, the whale eye is often described as 'avoidance'. It is used when a dog wants to avoid direct eye contact or break direct eye contact, in other words, the dog wants to look away from the problem without moving his head. In a highly tense situation, the slightest movement of the head could be misinterpreted and provoke an attack.

If a dogs fails to look at you directly and, instead, looks from the side, showing the white of his eyes, it could a sign of upcoming aggression if accompanied by other signals which support this premise.

A whale eye may also be visible when a dog moves his head, but his eyes don't follow the movement, for example during play.

A dog will often show a whale eye when he is guarding a bone, a toy or other resources. The white sclera also becomes visible in dogs with very tense, rounded eyes.

IDENTIFYING WHALE EYE

This dog is concerned about the hand that is coming from above his head. Many dogs find contact from above uncomfortable, although many guardians are unaware of this sensitivity.

White of the eye visible – tense

- Slit-shaped eyes
- Looking upwards

Additional clues indicating tension:

- Ears held towards the back
- Corners of the mouth pulled back
- Tightened facial skin

The loving hug is proving a bit much for Chemsi. She leans away and looks in the other direction.

White of the eye visible – slight tension

- Lightly rounded eye
- Eyebrows raised slightly

Additional clues indicating slight tension:

- Ears pulled back slightly
- Mouth closed
- Point of balance pushed to the side
- Right paw tensed

WHITE OF THE EYE VISIBLE

Attentive – guarding – warning

Toy guarding is usually very subtle and, generally, direct eye contact is enough to keep others at bay. Growling is rarely necessary. However, there is also a stillness that can be observed. It is important to be mindful in these situations, especially if children are present, as they will be unable to interpret the dog's emotional state.

WHITE OF THE EYE VISIBLE

Indignant to insecure

Whale eyes are often seen when the dog is feeling insecure and would like to turn his head away, but doesn't dare to take his eye off his opponent.

Jake, a four-month-old Border Collie puppy (*right*) is being very pushy with Luke, who is clearly unhappy. He shows whale eyes, which is the result of turning his head and trying to keep Jake in his field of vision. He also shifts his body weight back and away from the puppy, and pulls his ears flat and back. The area around his mouth is tight.

You can also see that Jake, the puppy, is experiencing a conflict of emotions. His eye is rounded and the whites of his eyes are clearly visible. His ears are pointed forward, but the corners of his mouth are pulled back and down. He would love to make more contact with the older dog, and can't quite understand Luke's reluctance to play. This is a very typical situation that occurs when young dogs are attempting to interact with older dogs. It is annoying to older dogs, and younger ones find it hard to understand. *Learn more about Luke and Jake's relationship, see The ears, page 42 and Tongue flick and air licking, page 259.*

WHITE OF THE EYE VISIBLE WHILE PLAYING

You will often see whale eyes when two dogs are playing together. This is usually because their eyes are open wide, and they are looking around a corner to keep an eye on each other. In both photos, we can tell the dogs are playing by their loosely flying ears and relaxed mouths.

Unusual eyes, eyelids with pigmentation, heterochromia (where each eye has a different coloured iris), and eye markings, don't express emotions, as such, because they do not change and the dog has no control over them. However, they can accentuate the emotion that is being expressed. This can create a false illusion of an emotion, or make the emotion seem stronger than intended. For example, direct eye contact can be misconstrued as staring, or a slightly raised eyebrow may be mistaken for one raised in anger.

Some dogs, like Joey, have such large eye openings that there is always a bit of white visible, even when he is relaxed. This can cause other dogs to think he is insecure because his neutral eye is difficult to recognise.This could provoke an attack.

Excessive pigmentation on the eyelids, and a lightly-coloured iris, can result in a stare-like gaze, which irritates many dogs. This husky appears to be very intense and fixated, but he is showing through his ear carriage that he is in a friendly, interested mood.

Dora, a Bavarian Mountain Dog, is a little insecure when she meets other dogs. However, in this photo she is expressing curiosity, but her light irises, accentuated by the dark markings around her eyes, cause other dogs to feel insecure.

HETEROCHROMIA AND FACIAL MARKINGS

Some breeds and cross breeds have differently coloured eyes, which is known as heterochromia. If other dogs are not familiar with this characteristic, it can be confusing and lead to misunderstandings. The same applies to dogs with very light-coloured irises, which are often combined with unique facial markings. Many dogs react to these unusual eyes by feeling insecure and withdrawing, others feel provoked and go on the offensive.

Dogs with different coloured irises (heterochromia) and irregular facial markings.

Dogs with light blue irises and irregular facial markings.

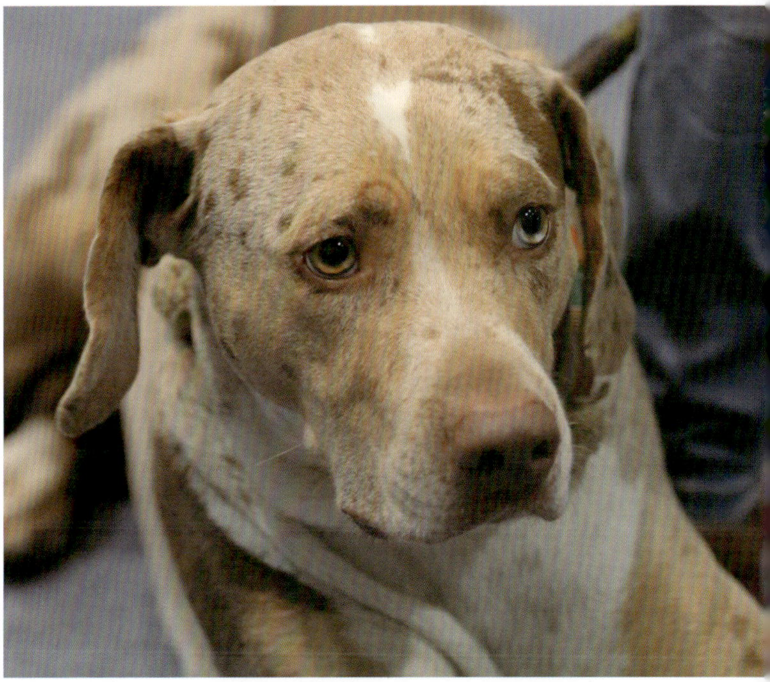

Dogs with amber irises.

EYEBROW MARKINGS

Markings around and above the eye accentuate the positioning of the eyebrows, and the edges of the eyelid, and, therefore, the facial expression. This can have a major impact on how a dog with these markings is perceived by other dogs, as he may appear to be far more intense than he intends. Some emotions will be displayed more clearly, while others can be easily misunderstood. Direct eye contact that is intended to show interest can be seen as threatening, or provoking, if a dog does not have experience with different dogs.

Litla, the Wolfspitz, has 'spectacle' markings around her eyes. By nature she is defensive and insecure, but her markings make quite a formidable impression on other dogs.

A face that is very dark, without markings, can lead to confusion because it is difficult to read.

Mixed breed, Oskar, is a very calm and confident dog, but as soon as he makes eye contact with another dog – without tightening his facial muscles – he is liable to be misread. Many dogs show concern and try to calm him.

IRREGULAR MARBLING (BRINDLE)

A marbled pattern on a brindle dog can draw the eye to specific features and affect the facial expression. This Sloughi appears to be raising an eyebrow.

PROMINENT FACE MARKINGS

Prominent face markings can lead to misunderstandings between dogs.

Gabi and Quivive met Cora, a female mixed breed, on a campground in France. She was tied to her guardian's camper and watched him walk away with concentration and some concern. Quivive immediately gave her extra space and walked past her in a large arc.

Signs of tension

- Eyes slightly rounded
- Staring

Additional clues indicating tension:

- Ears flat at the base and pulled back
- Top of the head is tight
- Head lowered
- Point of balance pushed forward

When her guardian returns, Cora is noticeably more relaxed. She now makes direct eye contact with the other dog; she has a much softer expression and is also blinking. However, her facial markings have made an impression on the other dog, which continues to avoid her.

More relaxed

- Eyes not as round
- Direct eye contact
- Blinking

Additional clues indicating relaxation:

- Ears pointing forward at the base
- No tension on top of the head
- Head raised and soft through the poll (between the ears)

IMPAIRMENTS TO THE EYE

Tearing, conjunctivitis, wrinkles

In some breeds, physical characteristics, such as the shape of the skull, the make-up of the eye, or excessive wrinkling, affect facial expression and can confuse other dogs. For example:

- A dog with a small skull has limited space for the eye sockets. This encourages the eyes to protrude and can result in tearing.
- Eyelids that roll inwards, and ingrowing eyelashes, can cause abrasion and tears.
- A blocked tear duct results in stained fur around the inner corners of the eyes, which is clearly visible in light-coloured dogs.
- In some breeds, the conjunctiva (third eyelid) is always visible due to the lower eyelid hanging down. As a result the eye is constantly irritated, often resulting in chronic conjunctivitis. These dogs suffer from constant itching or burning and tearing of their eyes, which can greatly diminish their overall wellbeing. Many will become touch and light sensitive, and will also become tense because of the pain they are experiencing. Vision may also be impaired due to the constant flow of tears. This can lead to excessive eye contact, which can cause misunderstandings when directed at other dogs. Dogs with conjunctivitis will be considerably more scent orientated and use their noses to 'see'.

The red and swollen third eyelid is clearly visible in this Great Dane.

Some breeds, such as the Shar Pei (*left*) or the Bulldog (*right*) have excessive facial wrinkles, which greatly influence the expression emanating from their eyes. Many dogs find this confusing, or unnerving. Excessive wrinkles can also affect a dog's ability to see and sometimes a surgical procedure is necessary to free up their field of vision.

Did you know that dogs can also be cross-eyed?

This is what it looks like.

7. THE MOUTH

When observing the mouth and nose area, a familiarity with the individual dog's face, in a relaxed situation, will prove very helpful

Of course, a dog does not use his mouth to speak, as we do, but dogs do a lot more with their mouths than simply eat, drink and emit sounds. They use their lips and the corners of their mouth for non-verbal communication, without making a sound. The position of the corners of the mouth relay a lot of information about the dog's mood, and the same applies to the area around and below the nose, the nasal sponge. Careful observation of this area can help us to predict a dog's next move, and his intended course of action. In addition, the way the tongue is held inside the mouth gives us information about possible physical or psychological stress.

In long-haired and/or black dogs with black eyes, the mouth, nose and tongue often gives us the only, easily recognisable, information relating to their emotional state.

VISIBLE DETAILS OF THE MOUTH, TONGUE AND NOSE

Detailed here are the parts of the nose and mouth that are important for the interpretation of a dog's emotions.

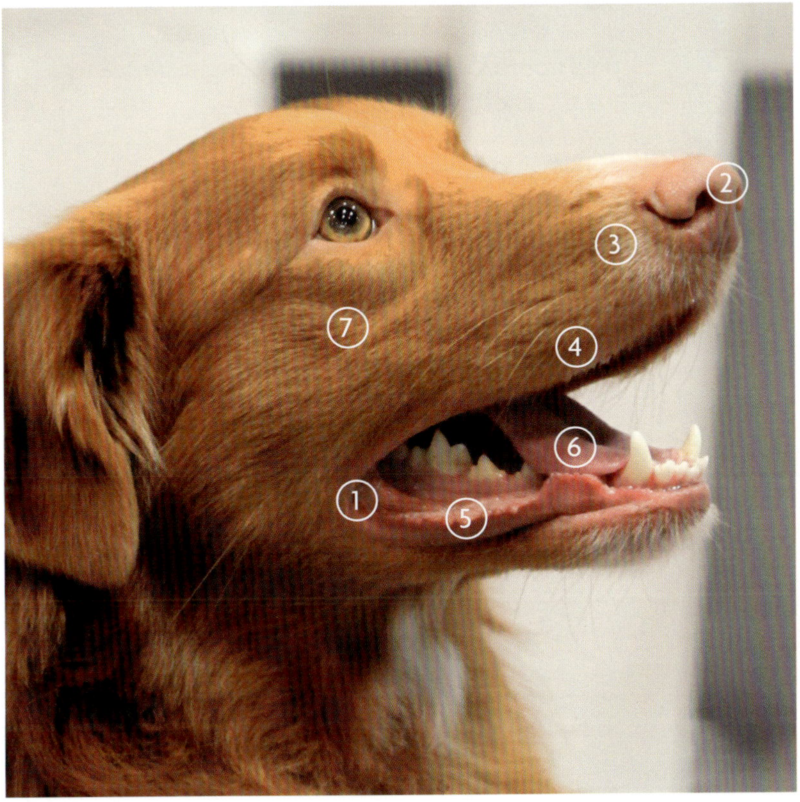

1. Length and shape of the mouth.
2. Nose and tip of the nose.
3. Facial skin around the nose (muzzle).
4. Upper lip with whiskers (vibrissae, *see page 225*).
5. Lower lip.
6. Tongue.
7. Cheek.

CLOSED MOUTH

The tightness of the skin around the face, or how much excess skin there is, affects the face when it is relaxed and when it is tense.

Very relaxed

- Mouth closed
- Upper lip covers the lower lip
- Hair smooth around the nose
- Nostrils normal size

Additional clues indicating relaxation:

- Eyes closed
- Ears hang down loosely

Here Quivive has a tense face, which is most noticeable around her mouth, with the tightened upper lip. This photo was taken at a campground during an approaching thunderstorm. Quivive was tired from a long car ride but could not relax due to the threat of the thunderstorm.

Tense

- Mouth closed
- Upper lip doesn't quite cover the lower lip
- Hairs around the nose raised slightly
- Nostrils widened slightly

Additional clues indicating tension:

- Eyes rounded
- Ears turned towards the back

Rani, a female Golden Retriever, has short facial hair, and a lot of loose skin around her nose and mouth. Here she is resting her head on the floor and is very relaxed. Her eyes are slightly open, and the skin around her mouth and nose flops loosely on the floor.

Relaxed

- Mouth closed
- Hair smooth around the nose
- Skin around the face hanging loosely

Additional clues indicating relaxation:

- Eyes slightly open
- Head resting fully on the floor

When Rani lifts her head, the wrinkles in her face change. Her ears are slightly raised at the base as if something has caught her interest. She appears to have frozen slightly, and is not fixating on a specific target. *See page 100 to find out what happened next.*

Very tense

- Mouth closed
- Mouth shortened
- Whiskers raised slightly
- Wrinkles developing around the nose
- Lower lip slightly visible

Additional clues indicating tension:

- Eyes more open
- Eyebrows raised slightly
- Head raised

CLOSING THE MOUTH AS TENSION RISES

When tension mounts, a dog will often close his mouth. This is generally seen in combination with a change in ear position and head carriage, wrinkles developing on the face, and the movement of the tail. By observing these signs, we can tell if the dog is:

- Paying attention
- Tense and cautious
- Tense but friendly
- Tense and annoyed

Maxi, a female Weimaraner, is watching a dog approach. As he comes closer, she becomes increasingly tense. The only evidence is that she has closed her mouth; her ear position and facial expression are virtually unchanged. The interaction with the approaching dog turned out to be peaceful.

In the photos this young dog is watching something with great concentration. By closing his mouth, his ear position also changes. He seems indecisive about how to respond to the situation. You can tell his concern by the change in his ear position and the lengthening of his mouth.

The ears are raised at the base and point forward.

The ears are turned to the side at the base and are only slightly forward. The mouth is more closed.

The ears are turned to the side at the base and held away from the head. The mouth is almost completely closed.

RESPONDING TO DIFFERENT EMOTIONAL SITUATIONS

A closed mouth is not always connected to increased tension. The dog may also choose to close his mouth when he is feeling relaxed, or when he is concentrating. You can evaluate the dog's emotional state by the tension on top of the head.

Rani, the Golden Retriever, is very relaxed. She looks at a dog, some distance away, with a soft facial expression.

Relaxed

- Upper lip covers lower lip
- No wrinkles around the nose

Additional clues indicating relaxation:
- Almond-shaped eye
- Soft eye
- Position of the eyebrows – neutral
- Ears hanging down loosely

Here, Rani has smelled something that catches her attention, and she is excited. She is concentrating hard. The skin around her closed mouth is tense.

Tense

- Slight development of wrinkles around the nose
- Upper lip does not cover the lower lip
- Corner of the mouth pulled slightly down

Additional clues indicating tension:
- Eye slightly slit shaped
- Eyes clearly focused
- Eyebrows pulled together slightly
- Head carried low
- Neck lowered

Dawina, the Dalmatian, is very tense and appears to be frozen. This is because she is receiving physiotherapy for cauda-equina-syndrome, which is a very painful condition. You can see a lot of tension below her eyes, and along the entire jawline. Note the detail in the photo inset.

Very tense

- Upper and lower lips meet and are joined tightly
- Corner of the mouth is pulled down
- Hairs below the bridge of the nose are raised
- Tight facial skin

Additional clues indicating tension:

- Large rounded eye
- White of the eye is visible
- Ears held towards the back
- Top of the head is tight

Here, Hadiya is clearly threatening Katja, who is standing beside her, taking photos. When a dog is making a subtle threat the mouth is usually closed, or just slightly opened, as opposed to an obvious threat where the dog exposes his teeth. However, Hadiya is serious in her message:"Don't come any closer!". *For more information on subtle threats, see page 97.*

Very tense

- Mouth almost closed
- Corners of the mouth shortened and pulled back
- Corners of the mouth pulled slightly down
- Whiskers raised

Additional clues indicating tension:

- Ears flattened and pulled back
- Eyes slit shaped
- White of the eye is visible
- Tight facial skin

OPEN MOUTH

Dogs can smile – or even laugh – and this is recognisable by the shape of the mouth. Similar to us, humans, the corners of the mouth are pulled up.

Even though the Greyhound has pulled his ears flat towards the back, he is enjoying the contact with Gabi who is standing behind him. His eyes are slit shaped, which could be the beginning of blinking. Many dogs will show this type of laughing face.

Relaxed

- Long mouth
- Corners of the mouth pulled up
- Bottom lip relaxed
- Bottom teeth visible
- Upper lip covers the teeth
- Tongue is relaxed inside the mouth

In this photo, we can see that Quivive, the Pyrenean Sheepdog, is in a really good mood. Her ears are pointing forward, her eyes are slightly rounded and she has an attentive look on her face. She is delighted with the spot she has found – seated on the lap one of her favourite people.

Relaxed

- Long mouth
- Corner of the mouth pulled up
- Lower lip relaxed
- Bottom teeth visible
- Upper lip covers the upper teeth
- Tongue is relaxed inside the mouth

Rani, pictured as a puppy, is happy in the midst of her family. Her face is relaxed, and her ears hang down loosely.

Relaxed

- Long mouth
- Corner of the mouth pulled up
- Lower lip relaxed
- Bottom teeth almost visible
- Upper lip covers the teeth
- Tongue is relaxed inside the mouth

Even when a dog is showing a laughing or smiling face, there can come a point where the contact, or petting, becomes too much. Happiness can quickly turn to stress. It is important to recognise the change in expression and respond if the dog is beginning to feel crowded. This is especially the case with sick or older dogs, as well as puppies. They need guardians who are aware of their emotional threshold, and who will take responsibility to resolve the situation.

SHAPE OF THE MOUTH

When the lips are pulled back, the expression is like a crooked smile which is mostly seen in dogs who are very tense. However, a crooked grin does not always indicate negative stress, it can also be shown by a dog who is feeling happy and excited.

If a dog is in deep concentration, he will usually close his mouth. If he doesn't understand a request, when training, he may indicate his confusion with a tongue flick, by blinking (or any other communication signal, *see Part II: Communication signals*), and a lengthened mouth. This tells us that we need to re-evaluate what we are asking the dog to do, and find a way to make the exercise easier so he can be successful.

This mouth shape can also be seen on a hot day, or if a dog is tired after vigorous playing. The shape of the tongue plays an important role in this situation and will give us a lot of additional information (*see page 105*).

CORNERS OF THE MOUTH PULLED BACK

The dog's breed, as well as the quality and quantity of skin around the mouth, can greatly influence the shape of the mouth. There are usually quite a few wrinkles behind the corners of their mouth, which can be tightened depending on the situation and degree of stress.

Quivive, the Pyrenean Sheepdog, is relaxed, but is panting slightly due to the heat.

Relaxed, mouth lengthened slightly due to the heat

- Corner of the lips pulled back
- Lips hang loosely
- Whiskers slightly raised

Additional clues indicating relaxation:

- Tips of the ears hang loosely
- Almond-shaped eyes
- Tongue rests loosely over the teeth
- No tension in the tip of the tongue

Spot, the Australian Shepherd, is tired following an intense play session.

Relaxed, mouth lengthened due to tiredness

- Mouth lengthened
- Lips hanging loosely
- Whiskers pulled back

Additional clues indicating relaxation:

- Ears hanging loosely
- Eyes almost closed
- Tongue resting loosely on top of front teeth
- No tension in the tip of the tongue

This German Shorthaired Pointer is feeling slightly tense. The corner of his mouth is comparably shorter and pointing down. The lower teeth are visible all the way to the premolars. His eye is not fully open, and the eyebrows are in a neutral position, which indicates slight tension.

Mouth lengthened with slight tension

- Mouth medium long
- Lower lip hanging
- Upper lip slightly tense
- Teeth visible
- Wrinkles around the mouth

Additional clues indicating light tension:

- Ear held slightly back

Milky, a female Bull Terrier cross, is relatively relaxed, which can be seen in her ear position, the soft, blinking eye and relaxed skin on her face. However her raised eyebrow, and a slight wrinkling of the skin, indicates a small degree of tension.

Mouth lengthened with slight tension

- Upper lip falling loosely and without wrinkling
- Whiskers not raised
- No wrinkling at the corners of the mouth

Additional clues indicating slight tension:

- Eyebrows raised slightly
- Ears raised at the bottom and turned to the side
- Wrinkles developing on the top of the head

Here, Milky is very excited at the start of a training session. The position of her ears, her raised whiskers and eyebrows tell us that she is tense.

Mouth lengthened with increased tension

- Mouth lengthened
- Slight wrinkles in the corner of her mouth

Additional clues indicating tension:

- Ears held flat towards the back
- Top of the head tight
- Eyebrows raised

This dog is panting due to the heat which is creating a lot of wrinkles and a lengthened mouth. Since he has a soft lower lip, with a lot of extra skin, his lip flops down and exposes the lower teeth, even though there is some slight tension.

Mouth lengthened with slight tension

- Mouth lengthened
- Lips hanging loosely
- Teeth visible
- Wrinkles at the corner of the mouth

Additional clues indicating light tension:

- Ears turned toward the back
- Eyebrows raised slightly

TENSION TO RELAXATION

These photos chart the early stages of tension, followed by relaxation.

From left to right: Looking at each photo in turn, we can see the tension rising in this light-coloured female – most particularly in the development of wrinkles at the corner of her mouth – until she gives a big yawn, and then relaxes.

This sequence was triggered when other dogs came a little too close. The light-coloured female didn't want to get up, but their proximity made her feel uncomfortable. The yawn indicates her indecisiveness.

These dogs are in a German shelter where they live together in groups and have many hours of supervised outside play. The play interaction between the light-coloured and the black female changes when the black and tan male joins in. This altered the group dynamic, which proved to be too stressful for the light-coloured female so she takes off. The time span of the interaction was 15 minutes, it was hot, and the light-coloured female was clearly exhausted afterwards.

THE SHORT C-SHAPED MOUTH

The c-shaped mouth, with raised upper lip and exposed canine incisors, usually means that the dog is serious in his intent. A confident dog often shows his teeth to warn another dog to reconsider, and avoid a confrontation. If the other dog continues to approach, he would attack without hesitation.

The c-shaped mouth can also be observed if a dog is in great distress, is fearful or, sometimes, when dogs are playing. It is therefore very important to consider the context, along with all other communication signals, before reaching conclusions.

This Alaskan Malamute is not happy about the tight leash. His tongue is pulled back inside his mouth, and is not visible, which may be a clear sign that he is intending to bite. His hackles and back hairs are probably raised, although this is difficult to see in breeds with this coat type. He is showing his teeth and should not be approached as it could trigger an attack.

Extreme tension

- Short mouth wide open
- Whiskers pointing forward
- Upper lip raised
- Incisors and canines visible
- Tongue inside the mouth

Additional clues indicating extreme tension:

- Ears pointing forward
- Slit-shaped eyes
- Fixed eye contact
- Point of balance forward

Rani is a little bit annoyed by Arwen's puppy advances. However, she feels settled around the puppy and, due to her kind nature, would not hurt him.

Slightly annoyed but still relaxed

Rani, lying on her chest:

- C-shaped mouth
- Upper teeth exposed

Additional clues indicating relaxation:

- Ears hanging loosely
- Head slightly turned away
- No direct eye contact

Bear in mind: not every dog will tolerate these puppy advances, and may not respond so kindly!

Chemsi, the Sloughi, lives with Inca, the Dobermann. Both are females and after spaying, their relationship has become more peaceful. Chemsi is a confident and, sometimes, cheeky dog, who usually decides the level of play between the two of them. Here she clearly shows her teeth; she has wrinkles on her nose and (*pictured left*), she has a c-shaped mouth. Inca is relaxed and tolerates her friend's cheekiness. Her mouth is open and relaxed, even when Chemsi nips her.

MOUTH WRESTLING

The c-shaped and wide open, or long mouths, can often be seen when two dogs engage in mouth wrestling. As long as the dogs keep showing some de-escalating signals, from time to time, it is fine to sit back, watch and enjoy the interaction. Direct eye contact is usually avoided and the dogs turn their heads away from each other to confirm that their intentions are playful.

Dogs with misaligned teeth can often be misread by other dogs because their mouths look threatening when their teeth are exposed.

Fosco was born with a protruding upper jaw which required a lot of veterinary attention when he was just a few months old. Now he lives a normal life and his malformation is no longer an issue.

When Quivive, the Pyrenean Sheepdog was out walking, she came across Tank, a dog with a badly disfigured face. Quivive approached him with great caution. Tank has an inoperable cleft palate which causes his upper teeth to be visible even when his mouth is closed. He is a very nice dog with a lovely expression and, fortunately, he has not been attacked by other dogs in the past. Most are careful when they first approach him, but quickly discover his kind nature.

DID YOU KNOW THAT DOGS CAN SMILE?

This is what it looks like:

Here Eve smiles at her beloved guardian, Katja.

Smiling is often misinterpreted as threatening when people don't know the dog very well. To differentiate a smile from a threat, it is important to also consider the situation and the rest of the dog's body language. Although the dog shows his teeth when he is smiling, he doesn't wrinkle his nose. Also, while smiling the tongue is not pulled back which you would see if a dog is exposing his teeth to threaten.

Most dogs only show their top incisors, although it also works in reverse

This behaviour can be seen as an expression of the need to belong, such as when a dog is very excited to see a person he knows well. Recognition of this behaviour is not a new phenomenon. Naturalist, Charles Darwin, made reference to smiling as far back as 1872.

This dog does not have an overbite – he's smiling !

WARNING FACIAL EXPRESSIONS

From warning to threatening

When a dog gives a subtle warming, the changes are often hard to see – even for the most astute observer – because they are often displayed for a mere split second. A warning, or "stay back" look, can change in an instant, depending on the response of the dog or person who is approaching.

A dog will display these subtle warning signs when he is guarding his personal space; he doesn't want the dog, person (or anything else) to come closer. He may be protecting a person he cares about, his food, a toy, a favourite blanket, etc.

It is especially important to be aware of the facial expression when small children are present who may approach the dog while being completely unaware of the signals he is sending out. They simply don't understand the warning signs.

In relation to your own dog, it is advisable to find out what he treasures and feels the need to protect, and how much space he requires. This will help you to predict his likely cause of action and defuse the situation. In terms of management, don't leave food or toys in a room where children are present, don't allow the dog into the children's rooms, and don't allow the children access to the dog's resting space.

How much space and quiet time a dog needs depends on his physical and emotional well-being at the time – and this can change quickly. If the whole family treats the dog with respect, and are aware of his needs, there should be no problems in the home.

Of course it is also important to teach the dog alternate behaviours so he does not have to resort to subtle warnings. Tellington TTouch®, combined with clicker training, are good tools to create relaxation and build calm confidence.

When dealing with unfamiliar dogs, we can create a safer environment for ourselves, our children and our dogs, when we are able to read the animal, respect him and be mindful of his needs.

Subtle warnings can be expressed with both physical and vocal stillness, but quiet grumbling, growling, or emitting other sounds, are all possible. It all comes under the umbrella of communication and provides us with the opportunity to de-escalate the situation and avoid aggression.

A dog who is not allowed to warn, or growl, is likely to go straight to an attack and bite. Therefore, we should always allow a dog to growl and, at the same time, look for an alternative training method to deal positively with the source of his concerns.

The following pictures show examples of both subtle and aggressive warnings. As we have mentioned in the previous behaviour patterns, it is very important to keep the entire situation in mind before judging.

FROM SUBTLE WARNING TO OVERT THREAT

Taschka is threatening the younger dog, Arwen, who is out of shot, but approaching from the right. The dogs know each other well, but puppies and young dogs often need a more severe warning before they respond appropriately and retreat.

Taschka is about to warn. She has already put her ears back and her eye is focused.

The subtle warning has grown into a clear threat with exposed teeth.

AGGRESSIVE WARNING

Seen from the front

Hadiya's message is clear: "don't come any closer!" Here, she is warning a person she doesn't know. She is serious and has attacked in the past in this type of situation. Her point of balance is far back, possibly in preparation for lunging forward to attack. Her whiskers which are pointing straight out to the side – a sign of her high arousal.

Extreme tension

- Upper lip raised
- Short mouth
- Whiskers raised
- Skin in the cheek area very tight

Additional clues indicating severe tension:

- Ears pulled back and flat
- Slit-shaped eye
- Direct eye contact
- Point of balance point – back

SUBTLE WARNING

In long-haired dogs

When a dog has a lot of facial hair, it takes a bit of practice to learn how to recognise the tension around the mouth and subtle raising of the upper lip. If you look closely at the photo (*right*), you can see the canine incisor. The forward pointing ears could have been created through movement, the eye is not visible, but it is safe to assume that there is direct eye contact.

Tension

- Raised upper lip
- Canine visible

Additional clues indicating tension:

- Head slightly raised

SUBTLE WARNING

In short-haired dogs

Abaju, a male Rhodesian Ridgeback, is a very confident and emotionally well balanced dog. He is a little stressed during the photo shoot because there are a lot of dogs he does not know in the vicinity. He warns very subtly with a slight raising of his upper lip.

Very tense

- Mouth slightly open
- Upper lip raised
- Wrinkles around the top of the nose
- Lower incisors visible
- Mouth rather short

Additional clues indicating tension:

- Eyebrows pulled towards each other and raised
- Ears held towards the back and side

SUBTLE WARNING

Seen from the front, and side

This dog was photographed at a show, a busy environment with crowds of people, and dogs coming too close to each other. Unfortunately many dogs are exposed to this type of challenging environment and are unable to cope. Next time you are at a show, take a closer look at the dogs and you will see clear evidence of stress. Note in particular, a mouth that is nearly closed, wrinkles on top of the nose and fixed eye contact. The dog in this photo has one important message – “don’t come any closer!”

Tension

- Mouth nearly closed
- Lips shortened and pulled forward
- Light wrinkling on the bridge of the nose
- Slightly flared nostrils

Additional clues indicating stress:

- Eyes rounded and fixed
- Eyebrows slightly raised
- Pupils large
- Ears pulled back
- Top of the head is tight
- Head raised
- Neck raised

FROM TENSION AND SUBTLE WARNING TO RELAXATION

Rani, the Golden Retriever, is lying down during a training session. A toddler crawls past, and you can see that she dislikes the interaction. It could be that the toddler has started coming towards her. Here, she shows the early stages of a subtle warning, followed by relaxation.

Tension and subtle warning

- Mouth slightly open
- Upper lip lifted
- Whiskers raised
- Lower teeth visible
- Clear wrinkling around the bridge of the nose

Additional clues indicating tension:

- Noticeable raising of the eyebrows
- Eye rounded
- White of the eye visible

Noticeable relaxation

- Mouth closed
- Upper lip covers lower lip
- Whiskers barely raised
- Very small wrinkles on the nose

Additional clues indicating relaxation:

- Eyebrows in neutral position
- Almond-shaped eye

In this scenario Rani knows the toddler well, and the child usually keeps more of distance from her. The interaction was closely supervised by an adult. Rani was not tied up; she is a very confident, friendly dog and always has the choice to get up and leave if a situation becames too much for her. The very quick wrinkling of her upper lip was a momentary communication, showing that she was slightly unimpressed. However, unless you know the dog exceptionally well, it is always advisable to interrupt an interaction of this sort, and move the toddler out of the dog's space. Every dog can have a bad day. A stressful experience in the past, spatial, or physical limitations, could trigger a change in behaviour. These factors should always be always be taken into consideration when dogs and small children interact.

IT'S ALL ABOUT THE STUFFED TOY!

Taschka is playing with a stuffed toy. Rani, who is out of shot, approaches from the left (and is interested in the toy). Taschka turns her head away, but keeps an eye on Rani as she comes closer. The combination of Taschka's closed mouth, and her position right over top of the toy, clearly signals to anyone who wants to come close: Stay back!

Tension

- Mouth closed

Additional clues indicating tension:

- Turning the head away
- Stillness of the body

Toys, sticks, and even a hole a dog has dug, can trigger subtle warning behaviour. It simply depends on what the individual dog values.

Taschka now turns her head in Rani's direction and relaxes slightly, which we can see by her open mouth. It could be that Rani stopped her approach at that moment.

Right after Rani turns his head to the left, Taschka also turns her head to the left.

In response, Rani turns her head again. Taschka closes her mouth a little more as her tension rises.

Rani comes closer, so Taschka turns her head to look at her.

As Rani is now close to Taschka, she goes to step over the toy and remains still. Her head is dropped and her nose is pointing towards the treasured toy.

This is how she claims the prize for herself – a behaviour dogs will also display when interacting with humans.

SUBTLE, SERIOUS WARNING THAT DOESN'T CHANGE

This German Shorthaired Pointer warns the other dog, who is in front of him, not to come any closer. The signal of sucked-in lips is a subtle warning. He may also be growling at the same time.

Tension

- Short mouth
- Start of wrinkles on the top of the nose
- Lips pulled inwards
- Whiskers/Vibrissae standing up

Additional clues indicating tension:

- Eyebrows raised
- Eyes rounded
- Ears pulled back
- Top of the head tight
- Head stretched forward

The GSP stops moving, he freezes his body, as well as his facial expression, while the other dog walks past him.

The dog who walked past has now gone. The GSP dog keeps tracks of him as he turns his head away. This creates a whale eye.

SUBTLE COMMUNICATION

With mild tension

Rani (*sitting*) is being 'courted' by the puppy, Arwen, who, with a lot of physical contact, invites her to play. Rani tries her hardest to ignore the puppy's advances and appears impassive. This is a common strategy used between adults and puppies when the older dog wants to indicate that he is interested in this type of contact.

Minimal tension

- Mouth closed
- Skin on her face seems smooth (no wrinkles)

Additional clues indicating minimal tension:

- Eyebrows raised slightly and pulled together
- Ears held to the back and side

Additional clues for relaxation:

- Soft eye
- Looking off into the distance

9. THE TONGUE

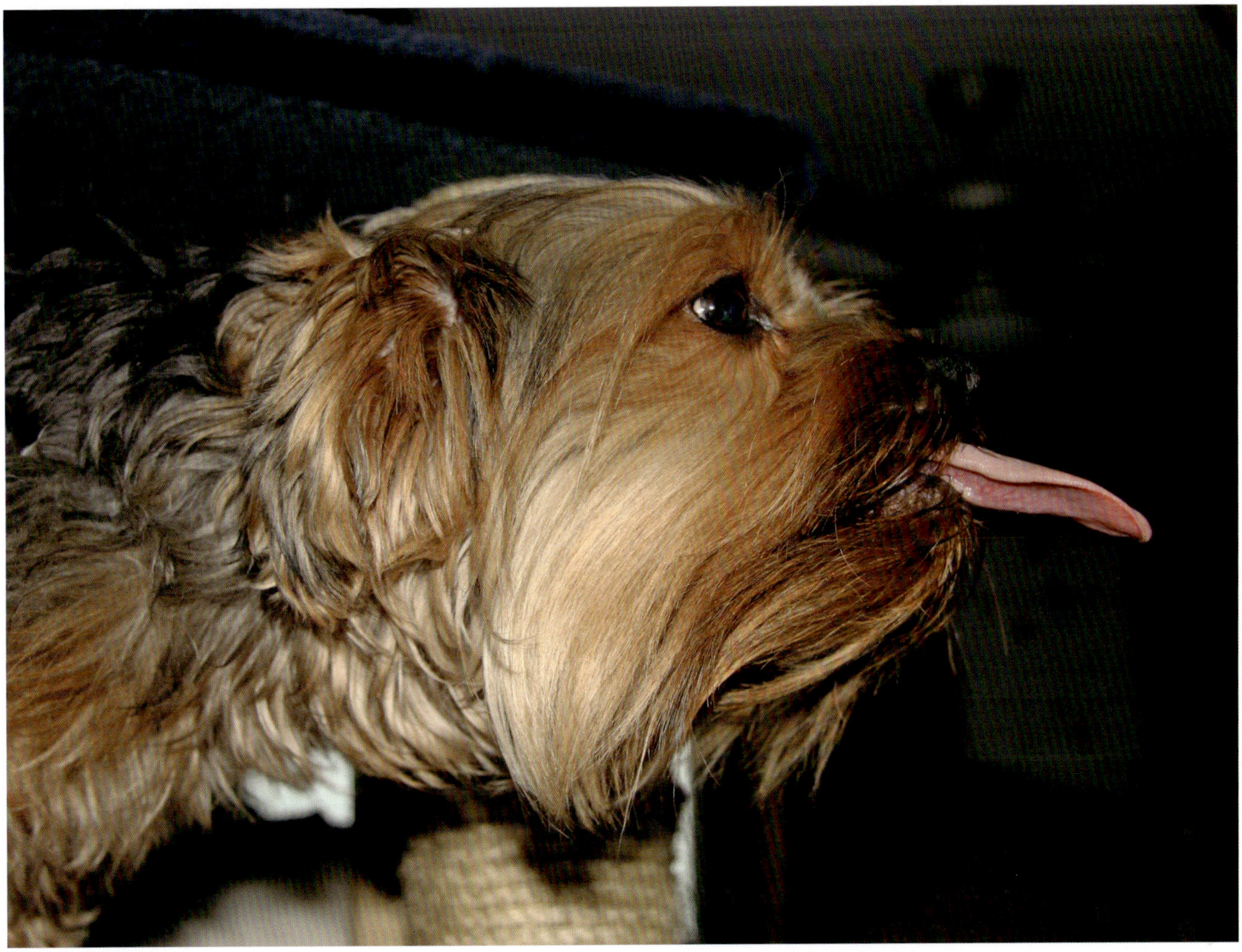

The tongue is a good indicator of a dog's physical and emotional well-being. A tongue that lies relaxed in the mouth gives us different information than one that is pulled far back, or hangs out of the mouth and is stretched to its capacity with a widened tip.

Dogs have their own individual tongue shape, length, width and colour. Some breeds, such as the Chow, are born with a blue tongue.

In the case of shock or poor circulation, a dog's tongue, as well as the mucous membrane, can change to pale pink, almost white in colour. When a dog is really excited, his tongue may become dark red or purple-blue. If the tongue starts to turn blue, with very little exercise, the dog may be experiencing a problem with his heart. Naturally pink mucous membranes will often become dark red if a dog gets excited, and the tongue is often affected by the colour change as well.

WHEN THE TEMPERATURE RISES

The tongue's most important function is regulating the dog's internal temperature. Therefore the dog is able to lengthen and widen his tongue in order to expand its surface area and speed up the cooling process.

As opposed to humans, dogs are barely able to sweat – they only have a few sweat glands between their toes – so they need to regulate their body temperature through panting, and passing air across the tongue.

Dogs do not tolerate heat very well so a rise of only one or two degrees Celsius (2 to 4 Fahrenheit) can be a matter of concern. Panting is an effective tool to cool the body, but it requires a lot of fluid. If the fluid expelled through panting is not replaced right away, the blood thickens and the body stops the panting activity to prevent more fluid loss. This will raise the dog's temperature and an increase to 40 degrees Celsius (104 Fahrenheit) can be life threatening. A dog that is locked in a car in the summer, even with the windows open, can reach that life threatening internal temperature very quickly. Opening the windows is not sufficient to cool the inside of the car; dogs still overheat.

It is also important to note that dogs also pant when they are in pain or experiencing anxiety.

PANTING IN HIGH TEMPERATURES

Rani's tongue while experiencing mild tension with lightly rolled edges near the tip...

... a little more relaxed, letting the tongue hang more, the tongue is spade shaped...

... even more relaxed, the edges of the tongue are smooth and flat, the tip wide.

Dogs with short noses pant rapidly when it is hot, or if they become excited, because their air and oxygen exchange, via the nose, is limited. To expand the area, the dog needs to stretch his tongue far out of his mouth. If air deficiency is still a problem, he will curl the outer edges of the tongue upwards.

When the dog starts curling his tongue, water, shade and rest are urgently needed.

TENSION DUE TO HEAT AND ANXIETY

Romeo, a mixed breed, lives in Tunisia and is chained up most of the time. Here he shows an extremely long tongue as a result of the heat. In addition, he is likely to be experiencing a certain amount of anxiety due to being photographed and the proximity of a stranger.

- Tongue hangs long from the mouth
- Slight spade-shaped thickening at the tip
- Edges of the tongue raised slightly
- Tongue muscles slightly raised

Additional clues indicating tension:

- Corners of the mouth pulled back
- Slight wrinkling in the corners of the mouth
- Whale eye (*left*)
- Ears held to the side and back

Large dogs also show a rolled tongue when their body temperature rises too much.

RELAXED TONGUE

This is what a relaxed tongue looks like. Rani's tongue sits loosely between her teeth and the tip of the tongue hangs out slightly.

CHANGES THROUGH TENSION

Tongues change their shape, colour and length during physical exertion or emotional excitement.

Typical of a dog experiencing stress: The tongueis sticking out of the mouth a long way and has a spade-shaped tip

This dog is not stressed; his tongue sticks out due to misaligned teeth. This is a common feature in many small dog breeds.

RELAXED TONGUE IN HEAT

- Rests loosely on the teeth
- Normal midline

The slight widening of the tip of the tongue is due to heat.

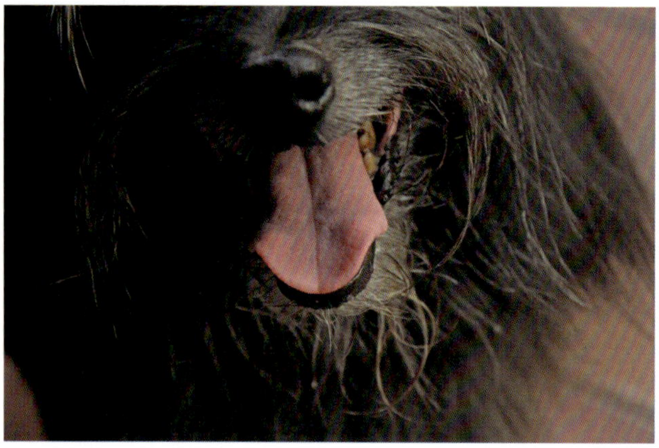

SLIGHTLY TENSED TONGUE IN HEAT

- Raised up slightly over the teeth
- Tongue muscles slightly raised
- Deep midline
- Foam developing

The tongue is tensed which causes the midline to become more pronounced, and the tip of the tongue to be spade shaped.

TENSION WITH CHANGE IN COLOUR

Due to excitement

Nash, a Nova Scotia Duck Tolling Retriever, has pink pigmentation. When he is excited his lips and eyelids change to a dark red. His nose, mucus membrane and tongue are also changing colour.

- Tongue held tight inside the mouth
- Tongue colour darkened
- Tongue muscle raised high
- Midline deep

Additional clues indicating tension:
- Corners of the mouth pulled back
- Eyebrows raised slightly
- Eye rounded slightly
- White of the eye visible
- Ears pulled back

TENSION WITH FOAMY SALIVA

On-going excitement

This female has been extremely agitated for the past half hour and is physically and emotionally exhausted. *Find out more about this situation, see Rage, page 440.*

- Long tongue hanging out of the mouth
- Foam developing on the tongue

Additional clues indicating tension:
- Corner of the lips pulled back
- Wrinkles at the corners of the lips
- Facial skin tight
- Ears pulled back and flat
- Top of the head is tight
- Eyes slit shaped

TENSION DUE TO PHYSICAL EXERTION AND EMOTIONAL STRESS

The little Pug is begging the rough, entire male, Staffordshire Bull Terrier Helge for a break. He has been harassing her and, in this case, the Pug's tongue is not just expressing her physical exhaustion, but also her emotional distress. If the Stafford keeps misinterpreting her signals, the dogs will need to be separated.

- Tongue extremely long and tongue muscle raised
- Tip of the tongue raised

Additional clues indicating tension:
- Mouth opened wide
- Corners of the mouth pulled back
- Ears flat and pulled back

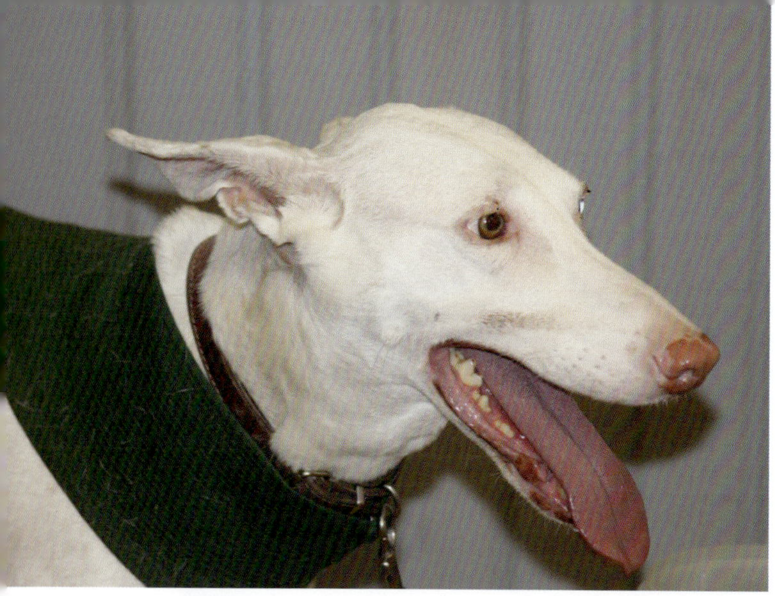

TENSION WITH CONTINUOUS FEAR

Diana is terrified of thunderstorms. The dark colouration on her tongue indicates a lack of oxygen due to excessive panting.

- Tongue long, hangs out of the mouth and is dark
- Midline groove deep

Additional clues indicating tension:
- Corners of the mouth pulled back
- Ears flat and pulled back
- Top of the head and skin around the face is tight
- Eye very round
- White of the eye visible

TENSION DUE TO PAIN

Quivive is experiencing severe physical stress as a result of excessive vomiting. She is panting hard due to pain and discomfort.

- Tongue lengthened
- Tip of the tongue widened
- Tongue held over the teeth
- Tongue muscle raised
- Midline of the tongue deep
- Edges curled upwards

Additional clues indicating tension:
- Ears held towards the back
- The eye we can see is rounded

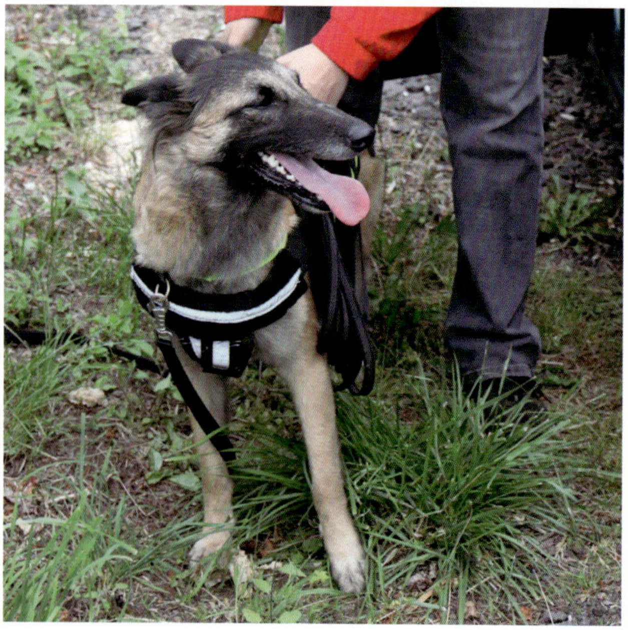

TENSION DURING TRAINING

Maja has been waiting in the car for a long time for her turn at a man-trailing training session. When her handler finally put the harness on her, she is very excited. She appears to be eager for her go, but we would question her stress levels while she is waiting. We would recommend some relaxing Tellington TTouches®, or some other form of calm exercise, to prevent her from wasting her energy unnecessarily.

- Long tongue hanging out of the mouth
- Tip of the tongue widened

Additional clues indicating tension:
- Corners of the mouth pulled back
- Ears flat and pulled back
- Top of the head is tight

EXCITEMENT

Positive stress due to excitement can turn into negative stress (distress) when the dog gets over-excited. The vegetative nervous system is activated due to the excessive dumping of stress hormones into the bloodstream, and the activation of the sympathetic nervous system. This is not a healthy situation for the body, especially if it is long lasting and repeated often. It diminishes the body's ability to relax and deal with stressful situations and increases the dog's tendency to overreact or become withdrawn.

The longer the dog stays in a negative stressful situation, the more physical symptoms develop. *See The five Fs, page 398.*

With this type of dog, it is important to create moments where the dog is happy but calm and quiet, making sure you stay calm yourself. This will help settle the dog and teach him to remain calm.

To monitor a dog's stress level, offer him a tasty treat. If he is too excited, he will either ignore the treat, take it but then drop it, or grab it so quickly that he may, accidentally, nip your hand. Whether the stress is positive or negative is irrelevant.

EXCITED AND HAPPY

Quivive always gets over-excited when her favourite people come to visit.

Without this information you would not be able to tell , from the photos, whether she is experiencing positive or negative stress.

- Tongue muscles raised
- Tip of the tongue over the teeth
- Edges of the tongue lightly rolled

Additional clues indicating tension:
- Lips pulled back, mouth horizontal
- Ears pulled towards the back
- Top of the head tight
- Whale eye (*see page 69*)

COMMUNICATION THROUGH THE TONGUE

A tongue flick or air/lip/nose licking is another communication signal dogs have in their repertoire and it is used in many different situations. This type of licking may be employed when the dog needs to calm himself, or calm whoever he is interacting with. A tongue flick can be a request for a break or indicate friendly intentions. It is a polite gesture and often used when another dog approaches too quickly. Even young puppies know to use their tongues for communication. *For more information, see Tongue flick and air licking, page 259.*

DO YOU KNOW HOW THE DOG USES HIS TONGUE TO DRINK?

The tongue is a very long muscle that is anchored to the bottom of the mouth and attached through a series of tendons. Most of the tongue is movable, especially the thin tip of the tongue and the edges of the tip of the tongue. This is the location of the supporting structure called the lyssa, which turns into a type of hose that dogs can move independently from the rest of the tongue.

When drinking (and eating) dogs can fold their tongue and use it like a ladle.

9. THE NOSE AND BREATHING

Unlike humans, dogs can process smells while inhaling and exhaling. They can differentiate between the right and left nostril, and have the ability to smell and analyse odours in specific areas of the nasal passageway

Sadly, short-nosed dogs (brachycephalic), for example, the Pug, and the Bulldog, often suffer from shortness of breath which may influence their health and quality of life. Their breath is usually raspy and they often emit a snoring sound, which can be irritating or confusing to other dogs.

The sense of smell is more important to a dog than their sight. A blind dog can still navigate his surroundings, with the help of his nose, to the point where guardians are unaware that their pet has lost his sight. Sensory organs begin to fail as the dog ages, and sight and hearing are usually the first to go. These dogs will start to spend more time scenting their surroundings, even if, previously, this did not seem so important to them.

As far back as 1938, Dr. Hans Georg Niemand wrote: "To imagine the olfactory abilities (of the dog) is only possible via a detour, by thinking of the dog's nose as directly comparable (analogous) to our eye."

The nose is the most important source of information gathering for dogs. It is, therefore, very important that they are able to spend sufficient time scenting during a walk and have the opportunity to analyse and explore a new area, with their nose, so they can feel safe.

A dog who is very excited, or insecure, may not scent when he comes to a new area because it requires concentration. He can only receive the odours when he is relaxed; stress blocks that ability and diminishes his awareness. Concentrated scenting and searching settles the dog and relaxes him, which improves his ability to smell and compartmentalise the information. It can, therefore, be a useful way of releasing tension when working with dogs who are nervous or insecure.

Dogs, like humans, store different odours in their long term memory. Odours are connected to emotions and different smells can trigger different behaviours in our dogs. We don't see many physical changes in the nose in emotionally charged situations, but due to the tremendous amount of information dogs receive through their noses, their emotions and behaviours may be influenced by the olfactory sensations they experience. We are only aware of a minute percentage of the smells our dogs sort through at every moment.

Dogs breathe in through the wing-like flaps in their nostrils, which open when they breathe in and then close again. Exhalation takes place via the lateral openings, the nasal slits.

Beagle, Lotte, is so captivated by her search that it appears she has been 'suctioned' to the carpet. The wrinkles around her nose show her intense level of concentration.

While Lotte wrinkles her nose and keeps the top of her head smooth, this Rhodesian Ridgeback wrinkles the entire top of hers, while her nose is relatively wrinkle-free.

This black-and-white dog keeps an eye on the photographer while sniffing the ground. Like the Rhodesian Ridgeback, he is not concentrating on finding odours; he scenting to calm himself.

UNDER THE MAGNIFYING GLASS

If you have a family pet, with strong hunting instincts, it is helpful to observe his nose, as well as the rest of his body, so you can potentially intervene when he catches the scent of prey.

Closed nostrils.

There is some tension in the nose and the nostrils are opened. The whiskers are pointing forward as the toddler approaches.

Nostrils wide open – Rani has caught the scent of ducks!

CHANGES AS THE DOG AGES

The tip of the dog's nose can change as he ages. Dark noses can lighten, pigment spots can appear, or the dog can develop small wart-like growths. If your dog's nose becomes very light in colour, it may become sensitive to sunlight and may require specialised sunscreen.

Rani (*pictured right*) at nine months.

At two years.

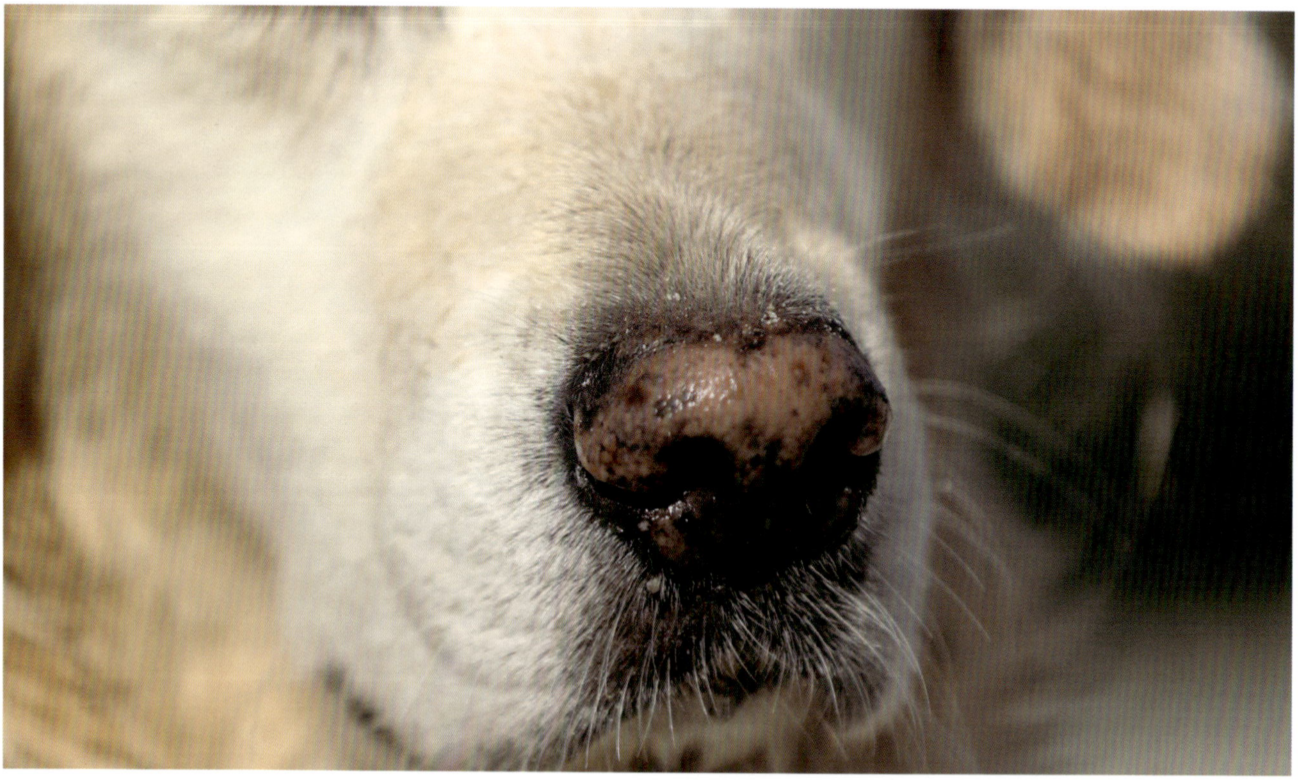

At 10 years of age. The colour on the tip of the nose can also change with the seasons; a dog with a dark nose in the summer may have a lighter nose in the winter.

THE JACOBSON ORGAN

The Jacobson organ (vomeronasal organ, VNO) is located in the dog's throat and has no access to the olfactory section of the nose. It receives its information exclusively through the nasal-palate channel from the oral cavity and the nasal airspace. The VNO mainly reacts to hormone-like messenger substances (fragrances or pheromones), such as those emitted while a female is in heat. It serves non-verbal and inter-species communication.

The processing of the information that the VNO receives takes place in the subconscious and has a direct effect on the emotions of the dog. When recording information through the VNO, various behaviors are shown, such as grinding of teeth, extreme saliva flow to foaming, loud lip smacking or licking the air.

This Weimaraner male is drooling and his saliva is foamy when he is near the female, who is probably in heat.

Guardians of entire males are likely to have observed this behaviour. When a dog is in this state he may not respond to verbal cues and should, therefore, be on a leash.

SECTION 2:

FOCUS ON THE BODY

1. PARTS OF THE BODY

When you take a look at your dog's overall body, there are different areas that will give us clear information about his emotional state. It is crucial to observe these body parts in a relaxed dog, i.e., in a natural or neutral posture. Many of the mood changes are indicated by very subtle changes in posture or body parts:

- How does your dog carry his **head** and how much does he raise his neck when he is in a neutral mood? The height he carries his head can also be seen by the height of the nose, while the raising of the neck is clearly recognisable by comparing it to the topline (back).
- What is the shape of the **topline**? It should be noted that different breeds have different anatomy and toplines which are not connected to the dog's mood.

1. Forehand. 2. Torso. 3. Hindquarters.

- The position of the **limbs,** and the angles of the **joints,** can give us a lot of information about how a dog is feeling. Again it is important to take into account that the dog's breed, his individual anatomy, and his age will influence certain positions of the limbs and angles of the joints.
- It is also helpful to observe where the dog maintains his **centre of balance.** Depending on the situation, a dog may shift forward or backward, and he may lean into or away from an object or situation.
- The **tail** is a highly visible mood barometer, which is held higher or lower than normal depending on the dog's state of mind, and the context. In an extreme situation, the tail will be tucked up, disappearing between the hind legs and clamped against the belly. The speed and eagerness with which the dog wags his tail also gives us a lot of information.

On an emotional level, the same, or very similar, body positions can have different and very individual meanings. It is, therefore, important to re-evaluate the dog as a situation changes.

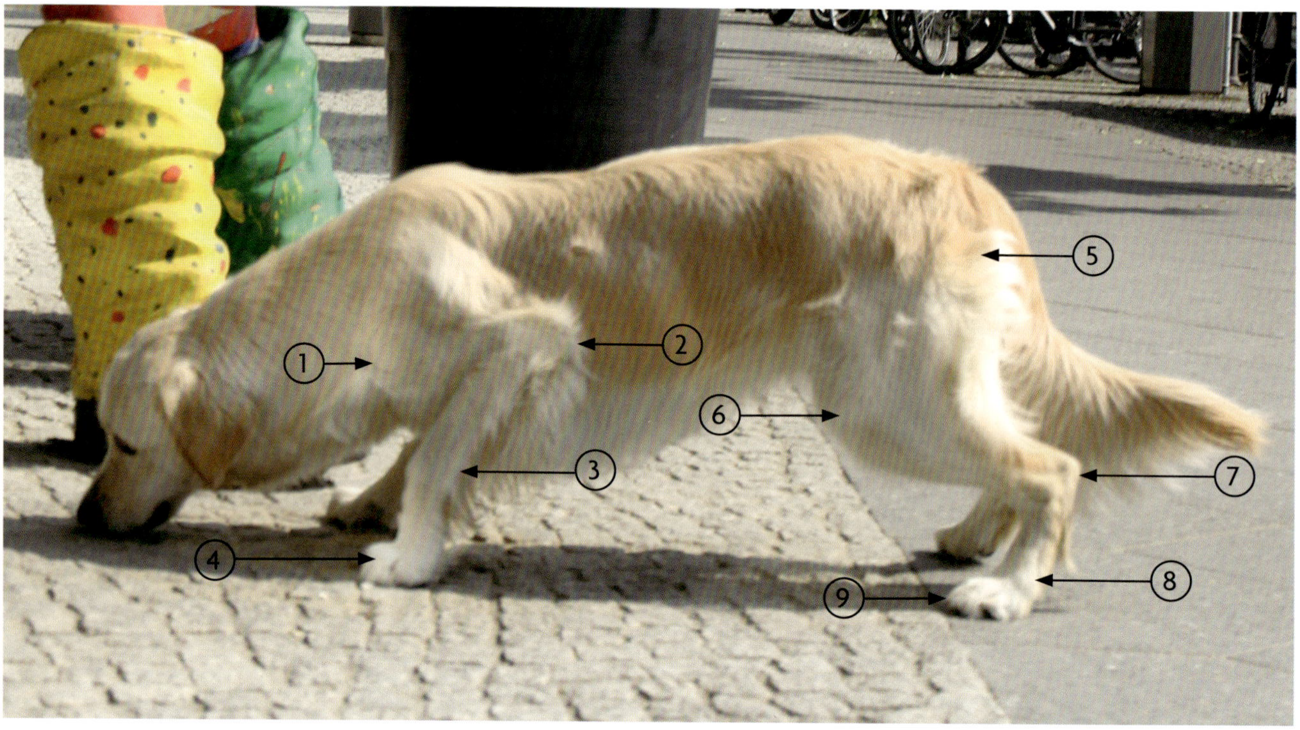

The forehand, or forequarters, includes the following joints:
1. Shoulder.
2. Elbow.
3. Wrist.
4. Toes.

The hindquarters includes the following joints:
5. Hip joint.
6. Knee (stifle).
7. Tarsus.
8. Metatarsus.
9. Toes.

COMPARING TOPLINES

There are significant physical differences between the toplines of various breeds, but there can also be a large variety of toplines within a breed. A typical example is the German Shepherd Dog (working and show lines) which shows the two extremes: from a straight to a sloping hind end.

Dodger, a Shepherd cross, has an almost straight topline; her hind legs are placed well underneath her body. When the back is very straight the middle of it can seem to be slightly dropped when the dog is relaxed.

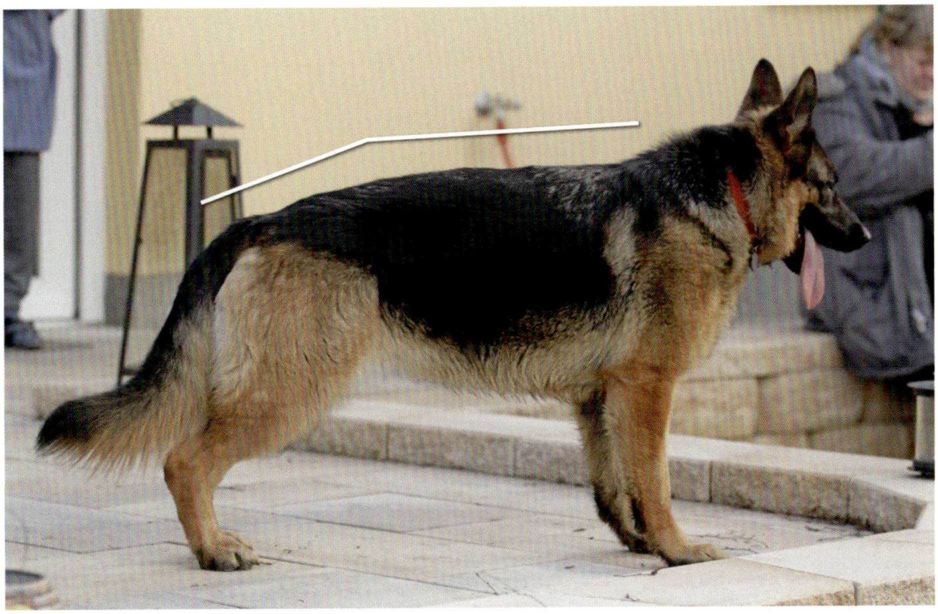

Püppi, a purebred German Shepherd, has a sharp drop to her sloping hind end; her hind legs are placed far behind her body.

A Greyhound with a slightly rounded back which is typical for the breed.

This Weimaraner has a slightly dropped or swayed back, probably due to his advanced age

COMPARING ANGLES OF THE JOINTS

The angles of the joints play an important role in the physical appearance of a dog. They are anatomically defined and change during movement, but also in emotionally charged situations such as anxiety or while 'posturing'.

Dhanyi , the Golden Retriever, shows strong angulation of her hind end. There is evidence of slight tension as she is focusing.

The angles of the joints are steeper in this Dalmatian as he is showing both focus and tension.

Angles of the joints in an English Bulldog when relaxed.

Try drawing your own lines to differentiate the body parts and angles of this dog. Good luck!

Extreme flexion of the tarsus in a German Shepherd with an extremely sloping back.

Within some breeds, such as the German Shepherd, we can find extremely different angulation of the joints due to opposing breeding goals (working and show lines)

Dogs with extreme angulation of the joints, like the German Shepherd (*above*), are generally not as balanced and secure on their feet, which creates emotional instability. Sadly they often suffer from arthritis and pain at a fairly young age.

Drigon, a German Shepherd, showing strong angulation of the hind legs and a downward sloping topline.

EXAMPLES OF POSTURE IN ONE DOG

A dog will use his body to express how he is feeling, so we need to become expert at interpreting the postures he displays. To illustrate this, we have chosen Quivive, the Pyrenean Sheepdog, as our subject.

Guardians of long-haired dogs tell us, again and again, that it is easy to read a short-haired dog and that long-haired dogs hide a lot of information. So we figured that if you are able to read a long-haired, black dog, you will have done all the hard work and will find reading short-haired, light-coloured dogs comparatively easy.

Quivive is black and long-haired and, therefore, she is not always easy to read. The position of her ears and the shape of her eyes can be hard to see, due to the long hair around her head. However, her tail position, the height of her head, her mouth, her point of balance, and the shape of her topline give us a lot of valuable information.

When Quivive is relaxed, her back is almost straight (slightly sloping towards the tail); when she is tense, it rounds a little towards her hind end. Her tail is set relatively high and the curled tip falls slightly to the side. When she is very relaxed, she carries her tail low with a slight curve at the tip. Her neck is slightly raised, and her head slightly lowered.

Quivive's joints show moderate angulation when she is relaxed, and standing, which corresponds to her natural/normal joint position. She is balanced evenly over all four feet.

Quivive has semi-erect ears. In their natural position, the ears are raised to about halfway up, and the tip of the ear hangs very loosely towards the front.

In the following descriptions, we are including Quivive's ear and tail positions, as we want to give a complete picture of what her body looks like.

STANDING STILL

When you are familiar with a dog's neutral, relaxed position – his head and tail carriage, and general posture – you will be able to notice the subtle changes as the dog starts to get tense.

Without tension during a walk

- Tail carried level with the back
- Tip of the tail falls to the side, and is moving
- Topline almost straight
- Head and neck raised
- Ears in natural position

Light tension in neutral position

- Tail carried higher than back
- Tip of the tail falls to the side
- Topline slightly raised towards the tail
- Neck raised, head raised higher
- Mouth open
- Joints slightly braced
- Balance point neutral over four feet
- Ears raised and turned to the front

Increased tension while playing

- Tail held even higher above the topline
- Tip of the tail raised, moving
- Topline rounded from front to back
- Neck and head raised
- More angulation to the joints in the hind end

Tension while playing

- Tail carried above the topline
- Topline rounded towards the tail
- Neck raised, head raised slightly
- Balance point moved to the back
- Ears held to the side

More tension while playing

- Tail carried above the topline and moving
- Topline almost straight
- Neck and head raised
- Slight bend in the joints
- Mouth open
- Balance point moved towards the back

Very high tension – hunting mode

- Tail lowered to be level with the topline – no movement
- Topline almost straight
- Neck high, head raised
- Right front paw lifted
- Ears raised and pointed forward to follow the line of vision

Gut feeling

If you have a dog that is a keen hunter, you need to be able to read his body language very quickly, and recognise the small changes in posture, so you can interrupt an upcoming chase. With practice you will develop that gut feeling. In these situations, some dogs may need to be exercised on a long line.

ON THE MOVE

It is even more difficult to pick up detailed information when a dog is moving. Photos help to educate the eye.

Walking – no tension

- Tail level with the topline, moving loosely
- Topline straight
- Neck level with the topline
- Balance point neutral
- Joints loosely bent

Walking – slight tension

- Tail carried above the topline
- Tip of the tail falls loosely
- Topline slightly rounded
- Head carried slightly higher than the topline
- Ground covering walk with loose, flexible joints

Walking – high tension (very insecure)

- Tail carried low, tucked at the base
- Topline falling forward
- Neck and head carried low
- Ears held flat and back
- Balance point moved forward
- Joints very flexed
- Hunched posture

Trotting – slight tension

- Tail held high above the back
- Tip of the tail falls loosely to the side
- Topline straight, falling forward
- Head carried slightly above the topline
- Dynamic forward movement with flexible joints

Trotting – increased tension (also stalking)

- Tail raised barely above the topline with little movement
- Topline almost straight
- Neck lowered
- Point of balance moved forward
- Dynamic, goal-oriented forward movement

Galloping – very little tension

- Tail raised
- Topline slightly rounded
- Neck and tail raised
- Average bending of joints

Fast gallop – tension

Quivive gallops very fast while carrying a stick. She balances through her head and tail.

- Neck above topline
- Topline slightly rounded
- Joints flexed

Top speed gallop – high tension

- Head almost the same level as the topline
- Topline very round
- Tail very long and straight behind
- Joints are extremely bent

TENSION WHEN INSECURE

It will be a great help to your dog if you can learn to recognise when he is feeling insecure so you are able to respond right away.

Increased tension with a tendency towards feeling insecure

- Tail raised high at the base
- Topline almost straight
- Neck raised high, head slightly raised
- Mouth closed
- Front legs – joints braced
- Balance point to the back
- Ears raised, held sideways
- Broad stance

High tension while very insecure

- Tail tightened at the base, lowered slightly
- Topline very rounded (roached back)
- Neck raised high, head dropped
- Mouth closed
- Front legs – joints braced
- Point of balance towards the back
- Ears pulled to the back

Extreme tension with severe fear

- Tail tucked between hind legs
- Topline very rounded, especially towards the back
- Neck and head raised
- Joints braced in the front
- Point of balance towards the back, leaning away from the other dog
- Ears pulled back

2. POSTURES

Depending on breed, and conformation, each dog has his own individual way of carrying his body, and the overall impression of the posture can vary. Try taking lots of photos of your own dog in a variety of situations so you become skilled at assessing his posture. This will, in turn, help you to evaluate other dogs.

Changes in posture and facial expression happen so quickly they catch us unaware. However, when you sit at home and watch it all unfold on the screen you will dive into a whole new world...

STANDING

Many herding dog breeds, such as Arwen (*below*) show increased angulation of the hind end and a more rounded back with lower tarsus. This posture, which is due to their anatomy, gives them the appearance of tension even though they are relaxed.

Relaxed

Arwen: Schafpudel (a traditional German shepherding breed).

- Tail held low, tip of the tail almost straight
- Topline raised towards the back
- Neck raised
- Head slightly lowered
- Ears hanging loosely
- Tarsus slightly bent

Tense

Neo: Australian Cattle dog.

- Tail hanging down loosely
- Topline lightly swayed
- Neck raised
- Head somewhat raised
- Ears held to the side
- Almond-shaped eyes
- No wrinkles on the face
- Not much bending in the joints

Increasing tension – while controlled

This male Rhodesian Ridgeback appears frozen between the two Weimaraner females.

- Topline rounded towards the back
- Base of the tail level with the back and still
- Joints braced and stiff
- Head raised slightly without movement
- Ears held towards the back

In this type of situation, which occurs quite often, it would be helpful to recall the 'controlling' dogs – the Weimaraners – using a friendly voice, because the Rhodesian Ridgeback is clearly stressed. Trying to recall the Ridgeback would not work because he feels he is not allowed to move until the two females 'release' him.

Tension with curiosity

Combined with caution

Tamika has seen something that caught her interest. She is curious, her balance point shifts towards the back, which indicates caution.

- Neck raised, head slightly raised
- Topline straight
- Tail held very high at the base and still
- Joints of the front legs are braced
- Balance point towards the back

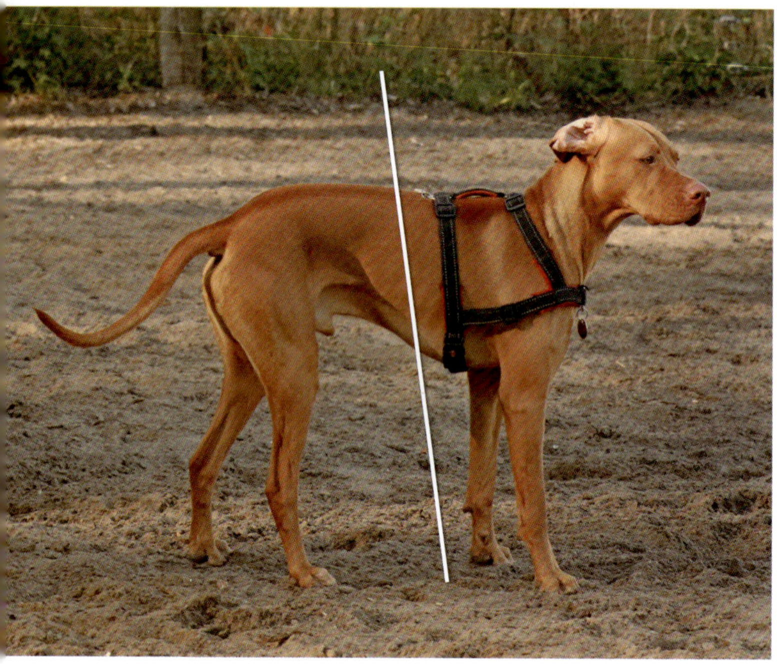

Raised tension

Increasing insecurity

- Back rounded towards the back
- Tail lowered and held tightly
- Angles of hind legs are very straight
- Joints of the front legs are braced
- Balance point towards the back
- Hair raised on top of the neck all the way to the tail
- Ears held towards the back
- Mouth closed
- Eyebrows raised

A harness that is too tight, or fits incorrectly, can greatly restrict a dog's movements and cause discomfort which will increase his insecurity. A harness with a strap that runs straight across the shoulders and the chest can negatively influence the dog's health.

If a dog pulls on the leash, a harness is a better choice than a collar, but the harness has to be chosen and fitted properly so that is not too tight, and does not inhibit the dog's movement.

Extreme tension with fear

The extremely rounded back, and tucked tail, show that the white dog is very insecure in this situation. She is being disciplined by the male who is entering from the left.

- Back extremely rounded
- Balance point towards the back
- Tail clamped
- Eyes round
- Ears turned towards the back

Extreme tension while insecure

The small, light brown female (*left*) is feeling very insecure. However instead of retreating, she barks at the Vizsla.

- Severe rounding of the back section of her back
- Tail level with the back and moving
- Balance point shifted back
- Crouching
- Severe bend in the joints
- Head raised
- Eyes focused on the other dog
- Ears pulled flat back

Extreme tension while insecure

Leading to threatening behaviour

Püppi, the German Shepherd, meets two, unfamiliar dogs (out of shot), and stands close to her guardian.

- Rounded topline
- Tail clamped at the base and moving
- Head at the same level as the back
- Ears towards the back
- Wrinkles around the nose
- Teeth exposed

TENSION WHILE GREETING

Many dogs greet the person they love, or dogs they are very familiar with, in a posture that looks like they are afraid that they will be beaten. They often show many calming signals at the same time, such as air-licking/a tongue flick, lifting a paw and turning the head away. Formerly this behaviour was labeled as 'submissive', and interpreted as the dog accepting his lower status in the social pecking order.

In his book, *Dog Sense*, animal behaviourist John Bradshaw gives this behaviour a different meaning. He sees it as being a gesture of belonging; belonging to a favourite person or a pack of dogs. In other words, it is an expression of love and a way to reaffirm the connection.

Increased tension when excited to meet a person

Here Quivive is greeting one of her favourite people. However, her posture, with her head and tail low, and the use of calming signals, gives an initial impression of a very unhappy dog.

- Tail lowered and moving
- Topline severely rounded
- Neck and head lowered
- Balance point towards the back

Sally, a small, mixed breed female, loves other dogs, especially Olli, a very charming Irish Wolfhound. She always greets him very enthusiastically. When looking at her signals, especially her crouching, the pinned ears and rounded back, you would think she is afraid, but they are all gestures of belonging. *See Joy, page 450.*

Increased tension when excited

Meeting a dog

- Tail held low but wagging freely
- Extremely rounded topline
- Head raised towards the other dog
- Ears held back, the left ear flipped over and relaxed

Dogs at play

Postural changes happen very quickly when dogs are moving, which makes it difficult to identify them. However, with a bit of skill and practice you can tell if dogs are happily playing as they are in the photo (*left*), or if the situation is serious and their guardians needs to intervene. The overall impression should tell us if the dogs are relaxed and moving loosely, if there is hesitation in their movement, or if the dogs are running towards a goal with fierce determination.

Arwen, the Schafpudel (*in the background*) is playing with Tamika. Even though there is quite a difference in size between these two dogs, who live together, their play is harmonious and well balanced. This is important when dogs of differing size and weight play together.

3. GAITS

Most dogs have three gaits: walk, trot and gallop. There are a few exceptions, such as the Kangal and the old English Sheepdog, which choose to pace as their middle gait.

The trot and pace are two-beat gaits, while the walk is a four-beat gait. The difference between them is that, when trotting, the diagonal legs touch the ground at the same time, whereas, when pacing, the lateral legs (the front right and hind right) are used simultaneously. A pacing dog gives the impression of rocking from side to side, similar to camels, giraffes and elephants. Some breeds of horse also choose the pace as a middle gait and are generally smoother to ride than horses that trot.

Many dogs will pace when they are walking on a short leash at a speed just a little too fast to walk and too slow to trot. They pace to adjust their speed to that of the person holding the leash. When following off leash, or on a long leash, most dogs will frequently switch between a walk and a trot. When a dog is old, sick or in pain, he will often choose the pace to protect himself. He can shuffle along without engaging and flexing his body as much as he would while trotting. Tired dogs also choose the pace because the gait expands less energy and is more shock absorbing.

Dogs with a lot of tension in their back, such as show dogs, may start to pace because they are constantly asked to raise and tip their head to maintain eye contact with their handler, which throws them off balance. This can cause arthritic damage to the vertebrae in the neck and also severe muscle pain in the neck and back. Emotionally stressed dogs, with a lot of tension in their back, can also show their discomfort by pacing.

The pace can be difficult to identify when a dog moves his legs in quick succession – taking short, hurried strides – in the walk. It is easier to recognise when a dog is trotting. In the gallop, watch out for a distinct hop, as the dog makes contact with the ground with both hind legs at the same time, to push thimself forward.

If your dog normally trots and gallops freely, and then suddenly starts to pace, showing an awkward gallop, take note and contact your veterinarian who can investigate the cause. A dramatic physical change such as this, when related to pain or an illness, will, inevitably, influence a dog's emotional wellbeing and may lead to behavioural problems.

THE WALK

The walk is a four-beat gait where three legs make contact with the ground at any given time. The footfall is hind left – front left – hind right – front right.

Relaxed walk

Weimaraner: Deluxe.

- Back almost straight
- Tail held below the back, swinging loosely
- Head held slightly above level of the back
- Ears hanging loosely to the side
- Eye almost closed

Attentive walk with sight tension

Australian Cattle Dog: Neo

- Back straight
- Head held slightly above the level of the back
- Tail held just below the level of the back
- Ground covering steps

Slight tension in the walk

Goal oriented

- Topline straight
- Tail carried higher than back
- Head low
- Clear extension of legs
- Ears pointed forward softly

THE TROT

The trot is a two-beat gait where the diagonal legs touch the ground simultaneously.

Relaxed trot

- Slight sway back
- Tail swings loosely with the rhythm of the trot
- Head held higher than the back
- Ears pointing back (slightly)
- Joints slightly bent

Slight tension in the walk

Feeling insecure

- Back slightly rounded towards the rear
- Tail slightly lowered
- Ears pointing back (loosely)
- Mouth open
- Neck raised

Slight tension with increased focus

- Topline slightly rounded
- Tail raised high
- Hair on back slightly raised
- Direct eye contact
- Head slightly raised
- Ears held differently from each other

THE PACE

When a dog is pacing, the legs on the same side touch the ground simultaneously. The pace, like the trot, is a two-beat gait.

Relaxed pace

Here you can clearly see the lateral movement of the right front and hind legs.

- Topline very tight
- Head and tail carried above the back
- Ears hanging loosely
- Eyes not focused on anything

Resembling an amble

Probably due to age

- Back swayed
- Tail clearly carried above the back
- Head and neck very low

Pacing, probably due to age

The lateral movement of the dog's left front and hind legs are clearly visible.

- Back straight
- Tail carried low
- Head and neck above the topline
- Ears raised loosely

THE GALLOP

The gallop is a jumped, three-beat gait. At a very high speed it may become a four-beat gait.

High speed gallop

Chase games at high speed, especially when the dogs are of very different sizes, need to be carefully monitored.

Relaxed gallop

Arwen, the Schafpudel, sees other dogs playing in the water and wants to join in.

- Topline straight
- Tail level with the back
- Neck slightly raised
- Head carried a little lower

Increased speed

They are playing chase, which is why Neo is holding his tail so close to his body.

- Topline rounded
- Tail clamped to the body
- Neck raised
- Head held towards the back
- Ears held towards the back
- Long mouth

Older dogs

As a dog ages, he may gallop less often, or not at all, as his spine becomes less flexible and his joints stiffen. Painful and laboured breathing will also be a hindrance when exercising at speed.

STALKING

It is easy to imagine that most dogs would feel threatened if another dog approached them in this stalking posture and the 'target' could easily become defensive. This can cause deep confusion in the 'stalker' who may misinterpret meeting other dogs with being attacked. Depending on their personality, dogs who have been bitten in the past will either become fearful or deal with future dog encounters with a strategy of: *attack is the best defence*.

We don't have any photos of these situations, because de-escalating the problem and avoiding a fight takes priority over getting a good picture.

Raised tension while stalking

When a dog approaches other dogs or people in this stalking posture, it is very impolite and can easily lead to misunderstandings.

- Topline straight
- Neckline dropping towards the head
- Head carried low
- Tail higher than topline
- Ears pointing forward
- Dynamic forward movement
- Point of balance forward
- Crouched posture

Appears to be stalking

Some in-bred, physical characteristics make it look as though a dog is stalking. For example, this German Shepherd appears to be stalking but, in fact, he is relaxed and is moving in a way that is perfectly normal for him. He is crouching due to his conformation and is unable to walk with a level back.

- Topline severely sloping down at the back
- Head carried a little lower
- Tail carried lower than the back
- Hunched posture
- Pacing
- Severe angulation in the joints of the hind legs

The exaggerated slope of the hindquarters has been seen as a desirable trait in the German Shepherd, and many breeders been selectively bred to for it. This breeding goal is very clear to see in the dog (*pictured above*). It is now known that this exaggerated characteristic can result in severe hip dysplasia.

POSTURE WHEN TIRED

Fatigue can be identified when muscle tension (muscle tone) lessens. When a dog is really tired, you will often notice a swayed back or neck. He will usually carry his head lower and the ears are no longer pointing forward, as he is losing interest. His eyes will also show a lack of focus.

Slight tension while tired

Here is Quivive at the end of a long walk in a field. She is tired from running and playing, but you can also see evidence of slight tension. If a rabbit crossed her path, she would immediately mobilise her energy and chase after it.

- Topline lightly rounded
- Tail lowered at the base
- Neckline swayed
- Head carried above topline
- Balance point towards the front
- Ears hanging loosely to the side, not raised

No muscle tension due to fatigue

This dog is coming to the end of a man-trailing session. His fatigue is evident through his lowered head carriage and the wrinkles on his face around the corners of his mouth. His ears no longer point forward with interest, but are to the side.

- Head carried low
- Ears turned to the side
- Mouth long
- Wrinkles on his face
- Long tongue
- Crouched posture
- Severe angulation of elbows and tarsus

Minimal muscle tension

Lotti (*pictured left*) is tired following a fun play session. Her posture shows the lack of muscle tension.

- Neck level with the back
- Long tongue
- Tip of the tongue widened
- Corner of the lips pulled back
- Wrinkles in the corner of the mouth
- Eyes rounded
- Ears pulled back
- Topline dropped in the middle of the back

Exhausted

Arwen, the Schafpudel, has just had a hectic play session with Tamika, an older dog in great shape. Arwen is much younger and tires more quickly.

- Ears hanging
- Long tongue hanging out of the mouth
- Limited body tension

Young dogs

It is very easy for young dogs to overdo it when they are playing, especially in puppy play groups, or when playing with adults. It is therefore important to observe young dogs closely, and watch out for signs of fatigue. Encourage dogs to take a break while playing, and consider the distance when going for a walk to avoid fatigue. This also applies to older dogs, or those who are in poor shape, and may be experiencing discomfort or pain.

4. POSTURING

What does posturing mean? We have spent a lot of time discussing the term, and trying to work out how it manifests itself in dog behaviour. The dictionary term for posturing is: "behaviour that is intended to impress or mislead". Can we apply this human interpretation to our dogs? Does the term take on a different meaning with regard to the opposite sex? This is a fascinating area, opening up the possibility of new ideas, and evaluating behaviour from a different point of view.

Posturing is usually seen when entire males meet; it could also be described as imposing behaviour.

POSTURING ON MEETING

Neo, the Australian Cattle Dog (*left*) has met a male he is not familiar with during his walk. Both dogs present a very similar picture, slowly circling each other with braced joints, raised heads and stiff tails. The aim is to impress, and settle who has more authority without needing to escalate to a physical confrontation.

- Topline fairly straight
- Tail severely raised at the base
- Neck severely raised
- Direct, frontal eye contact
- Joints of the front legs are braced

It seems that authenticity, presence and mental strength outweigh size, because small dogs will often engage much larger dogs in a posturing stand-off. Posturing does not always prevent a physical confrontation, but it often serves to clarify the question of authority without serious injuries.

Posturing can also involve the dog becoming still or freezing. This is the first meeting between Quivive and a little black-and white female, called Lucy. You can see from the positioning of Quivive's hind feet that she does not change her position, even though Lucy walks around her and sniffs her hind end.

POSTURING BETWEEN THE SEXES

The posturing of the white male (wearing the collar) is ambivalent. His raised hair shows that he is clearly impressed by the 'girl power' that surrounds him.

- Topline straight
- Tail raised at the base, held high and curled over the back
- Neck raised
- Head pulled back slightly
- Ears raised at the base and turned to the side
- Front legs are braced
- Hair on the back is raised

Many males show a very similar posture when meeting female dogs. Through the addition of the flirting ear position (*see page 43*) the posturing turns into a request to be liked – "Hey, look at me, I'm awesome!"

AMBIVALENT POSTURING

Showing minor insecurity

A mixed breed male, Fionn (*left*) meets Fredi, a Landseer, for the first time. He uses his posture to keep her at a distance. His pulled back ears show us that he is a little insecure in this situation.

- Tail raised high, no movement
- Neck raised
- Direct eye contact
- Balance point towards the front
- Hair raised in the shoulder area

SHOWING MAJOR INSECURITY

In this photo, it is tempting to say that Micky (*right*) is posturing towards Daniel, a male Golden Retriever. However, on closer examination, you will notice that Micky is holding his ears to the side and back, which does not fit with this interpretation. In fact, he is pretending to be confident to conceal his insecurity.

- Topline almost straight
- Tail raised high, no movement
- Neck and head raised
- Direct eye contact
- Front legs stiff and braced
- Balance point towards the front

FEIGNED POSTURING

When Maxi, a female Weimaraner (*left*) and Quivive meet for the first time, it appears that both dogs are posturing. Maxi uses direct eye contact, but her eyes are almond shaped and relaxed. Quivive seems to be looking away slightly. Both dogs express their friendly intent through these subtle signals.

- Neck clearly raised in both dogs
- Joints are braced
- Maxi's tail is held level with her back
- Quivive's tail is carried above her back
- Both tails moving
- Maxi's ears are raised at the base and held to the side
- Quivive's ears are held towards the back

EXTERNAL FACTORS

This dog's appearance is influenced by the tight leash which gives the impression that he is posturing. This is commonly misinterpreted by other dogs.

- Topline almost straight, slightly sloping towards the back
- Tail raised
- Joints braced
- Hind legs behind the body
- Balance point forward
- Head held high
- Neck clearly raised

POSTURING TO INVITE PLAY

Play is really important for dogs – especially when they are young – as it is a means of learning social interaction skills. Many dogs well into advanced age still like to play, but this does depend on the individual.

Dogs that do not have a chance to play as puppies, or youngsters, can struggle later in life, failing to understand a play invitation from other dogs, and even interpreting it as a threat.

The typical play bow, with widely spread legs, clearly expresses a desire to play, even when there is no direct eye contact. However, many dogs will use more subtle postures and signals when they are inviting another dog to play. While engaging in play, many dogs will keep displaying exaggerated movements and gestures to re-affirm to their partner that this is a game and everyone is having fun.

Pawing, sniffing and other signals can be used to ask for a break. An invitation to play does not have to be accepted. It is often used to simply establish a friendly-peaceful contact.

Tail carriage can vary in height and movement from dog to dog.

Many dogs have a preference for a particular game, and may be selective as to the playmate they choose. For example, when Quivive was a puppy she loved running games and rarely got involved in wrestling games. If another dog failed to respond to her invitation to chase, and wanted a different type of game, she simply ended the interaction. Dogs that approached her, wanting to wrestle, were ignored. Quivive would turn her head and walk away. In adulthood, she only plays with dogs that were her playmates when she was a puppy; she is more reserved with later acquaintances.

SUBTLE INVITATION TO PLAY

This is the typical posture Quivive uses to invite dogs, or her favourite people, to play her favourite game – "catch me if you can!" Her front legs are far apart and she lowers her front end just enough to get the message across. She is not engaging in a full play-bow so that she can take off more quickly when her invitation has been accepted.

- Very slight lowering of front end
- Joints of the hind legs slightly bent
- Front legs far apart
- Ears loose and different from each other
- Direct eye contact
- Head turned sideways (slightly)
- Mouth open (she is laughing)

A REPERTOIRE OF SUBTLE INVITATIONS

When the other dog is showing signs of insecurity, Quivive uses more subtle ways to invite play. This generally involves turning the head and a sudden leap to the side. She does not engage in direct eye contact.

Quivive always showed great skill when she was communicating with insecure and fearful dogs, and was never pushy when she invited them to play. Most of them were quick to trust her, and accepted her invitation for a game of chase.

For more examples of play invitations, see Bowing, page 358.

PLAY STALKING

This game involves lying in wait, until the other dog is in the right position, and then pouncing either towards the dog, or taking off hoping to be chased. It is a game many dogs love. The crouching dog will build up a lot of tension in order to bounce on to his feet and dash away.

Arwen is play-stalking another dog. Her mouth is slightly open and it looks like she is air-licking.

- Neck level with her back
- Head stretched forward
- Ears pointed forward
- Eyes focused on her friend
- Balance point towards the front

Arwen gets up and is gaining speed, but she stays in a crouched posture.

- Neck lower than head
- Head stretched forward
- Ears and eyes pointing forward
- Balance point towards the front
- Joints very flexed
- Mouth more open

Arwen's head and neck are almost level with her back and her tail is raised. She is getting closer to her target.

INTIMIDATORY STALKING

There are dogs who use stalking as a means of intimidation, and this can easily instigate an attack. It can be hard to identify the dog's intention purely by his posture, as it is very similar to the stalking posture of a dog who wants to play.

Fosco's posture is typical for a working Border Collie. His point of balance is far forward; his ears and slightly raised hind end tell us that he is not afraid and is very focused.

Major tension

- Balance point far forward
- Head very low
- Slightly crouched posture
- Tail held close to the body

Arwen is stalking a strange dog that has stopped and remains frozen.

- Balance point in the direction she is moving

Additional clues indicating tension:

- Head dropped
- Tail carried below level of the back
- Goal oriented forward movement

In this situation, both dogs should be encouraged to break position, and change their focus. Positive methods, such as clicker training, can be used to teach a dog to orientate towards his guardian, who can then ask for a recall. If a dog has a tendency to stalk, he should not be brought into situations where he is likely to adopt this behaviour. If it does happen, it should be interrupted immediately as the increase in tension can easily trigger an attack.

POSTURES REFLECTING DIFFERENT MOODS

From tense to relaxed

Lotti is a female with extreme emotions (*see Fear, page 414*) which she shows clearly through her body language.

Very high tension

The calming signals she uses are air-licking and lifting her paw.

- Topline very rounded towards the back
- Tail clamped between her legs
- Head carried level with her back
- Ears pinned flat towards the back
- Balance point towards the back

Minimal tension

- Topline slightly rounded towards the back
- Tail held low and close to the body
- Head and neck raised and carried higher than the back
- Ears flattened and pointing back
- Mouth closed

Tired from playing

In this photo, you can see slight tension through Lotti's tightened tail. She usually holds her ears pointing towards the back, even when she is relaxed.

- Topline almost straight
- Tail close to the body, hook shaped upwards
- Head and neck raised
- Ears held towards the back, barely raised
- Mouth open

INTIMIDATED TO INDIGNANT

In this photo, a white female mixed breed is suddenly attacked by a small, brown-coloured dog. She is feels intimidated, as shown in her posture.

- Back-end of the back rounded down
- Balance point shifted to the back
- Ears flat and held back
- The eye we can see is closed

The little dog perseveres in his attack, and the white dog's facial expression changes, even though her tail is still tucked and her back rounded. She is indignant, which you can see in her ears and facial expression.

- Back very rounded
- Balance point very far back
- Tail tucked
- Posture hunched (crouching)
- Ears raised at the bottom and held to the side
- The eye we can see is open
- Wrinkles on the nose
- Whiskers raised

Now she has had enough! She raises her front end and clearly threatens with her teeth. Her balance point stays towards the back.

- Topline straight
- Balance point still towards the back
- Tail raised
- Ears raised at the base, pointing forward
- C-shaped, short mouth
- Teeth visible
- Eyes small

The white dog air-licks, probably to calm herself, as the smaller dog retreats.

The aftermath: A very lovely dog with a soft, loving aura who was able to control her change of emotions and did not stay annoyed or get confrontational.

STALKING AND BULLYING

The black male has his eye on a young Great Dane (out of shot). There is an element of stalking in his posture as he heads towards his target in an extended trot. While engaged in his confrontational and targeted approach, his topline is very straight. Note: there are children present.

- Tail carried high with very little movement
- Head slightly lowered
- Ears held forward
- Line of vision goal oriented
- Dynamic forward movement
- Joints tightened

Now he slows down. Note the increased bend of his joints – he is getting ready to charge.

- Tail carried very high
- Neck raised
- Ears raised towards the front
- Line of vision goal oriented
- Balance point shifted to the front
- Dynamic flexion of the joints

When a dog is this tense, an attack is likely to ensue and, therefore, an intervention is urgently required. This type of posture, combined with direct eye contact, can be described as stalking or bullying. The term is used to describe continuous harassing and pestering of another dog, which will, at best, cause considerable distress and, at worse, will result in combat. This interaction needs to be interrupted, not least because the children could be at risk..

Now in relaxed posture

This situation was resolved with an intervention. The black dog's guardian called him away and rewarded him with a tasty treat. The difference in his posture and body tension, which reflects his emotional state, is very clear to see. When a situation, such as this, unfolds, early intervention is essential both to alleviate the distress of the targeted dog and to stop matters escalating.

5. POINT OF BALANCE

Dogs communicate by shifting their point of balance. For most dogs, the priority is to express a need for distance, either by increasing or decreasing it. To achieve this, a dog will shift his balance – his bodyweight – towards the front, the back, or to either side, depending on the desired direction.

When dogs interact, leaning into another dog's space is generally viewed as impolite. However, we humans are guilty of doing this, often being unaware that we are invading the dog's personal space. By observing a dog's shift of balance, and the direction he favours, we can gain valuable information concerning his comfort level in any given situation. Sometimes the shift is very clear to see but, at other times, it is very subtle and takes a trained eye. The degree of weight shift depends on the individual dog, and the specific situation he is dealing with. It is also indicates what may be his next move.

CONFORMATION

The ability to shift weight – and what it looks like – is affected by a dog's breed type, and his individual conformation. For example, the balance point of a Bulldog is closer to his muscled front end, whereas a herding dog's balance point is centred more evenly between his front and hindquarters.

Solid dogs are usually less flexible and cannot shift their balance point as easily as breeds that are more agile, lighter and quicker. As a result, these heavier breeds appear more expressive in their movements as they use more basic motor functions. This can create misunderstandings for dogs that move with more subtlety (*see page 19*).

If you want to evaluate a dog's centre of balance so you can detect a weight shift, you first need to find where his neutral point of balance lies when he is relaxed and in balance.

Schafpudel: Very flexible.

Bull Terrier: Not as flexible.

Bulldog: Even less flexible

Quivive and Feivel (*pictured right*) are in perfect balance because they are both in full control of their balance point.

The vestibular organ, which is responsible for balance, is located in the inner ear and is also responsible for turning the head and moving it vertically. In some cases, the vestibular organ does not function properly and the dog may suffer from dizziness. This may be experienced when the dog is in a moving car or on raised agility equipment. The vestibular system, as well as proprioception, which is responsible for the body's awareness of where it is in space, belong to the internal sensory organs.

Dogs with poor spatial awareness are often also unbalanced and it is, therefore, more difficult for them to express their emotions through body language. They appear to use more basic motor functions and struggle to keep the distance requested by other dogs, or people. These failings can be seen in many breeds, and are also evident in specific growth phases of young dogs.

Ischka keeps her point of balance right over the top of the tipping point of the seesaw. Her body is perfectly balanced, which prevents the seesaw tipping. Ischka is wearing a Tellington Body Wrap which supports her spatial and body awareness.

Rani, the Golden Retriever, is not comfortable on the seesaw. Her ears are pointing towards the back, there are wrinkles around her mouth, and the skin under her eyes appears to be tightened. Her tail is pointing down and held close to her body. Her paws are spread. She is not in balance.

Nero, the Australian Cattle Dog, is a master of quick evasion. When there isn't enough time to move, he simply jumps over the 'obstacle'.

Dogs have the ability to maintain their balance even when running at fast speed on uneven terrain. They can quickly and easily evade any obstacle that is in their way or change direction while playing.

BALANCE POINT SHIFTS BACKWARDS

Dogs generally shift their weight to the back to protect or increase their distance. Depending on the situation, this can occur subtly or be very obvious. We humans show the same reaction when someone, or something, is too close to us.

Slightly insecure

While standing

This dog lives in a shelter. He is very careful when dealing with strangers and when this photo was taken, he was just stopping his forward movement.

- Subtle shift of balance point to the back

Additional clues indicating tension:
- Topline rounded towards the back
- Tail dropped and held tightly
- Ears held to the side
- Mouth closed
- Front legs braced

Slightly insecure

While sitting

Here, Rani, the Golden Retriever, is meeting other dogs in their home. There is not a lot of space, which causes the dogs to be too close together, and it is difficult for them to move away. Rani, a youngster and a visitor, is tense and unsure how to act as she doesn't want to provoke the other dogs in the security of their own home.

- Balance point shifted towards the back

Additional clues indicating tension:
- Neck raised and head slightly dropped
- Ears flattened
- Mouth closed
- Whiskers raised slightly
- No direct eye contact

These dogs have pushed their balance points away from each other. Their laughing faces, exaggerated movements, and fun tail expressions tell us that they are enjoying an exciting play session.

All-round tension

When the small, white-and-tan dog approaches, the black Labrador shifts his balance point towards the back. His head is slightly ducked, and his ears are lying flat. He carries his tail slightly above the level of his back and it is moving. He is making eye contact with the other dog.

The smaller dog shows a large, rounded eye and he pulls his ears back. His tail is a little below his back and his balance point shifted to the side, away from the Lab. Both dogs are expressing tension.

Expressing fear

Eddy (*front*) shows his insecurity when the black puppy sniffs his rear end.

- Balance point towards the back

Additional clues indicating tension:
- Rounded back
- Clear angulation of the hind legs
- Strong angle of the elbows
- Tail under his belly
- Mouth closed
- Whale eye

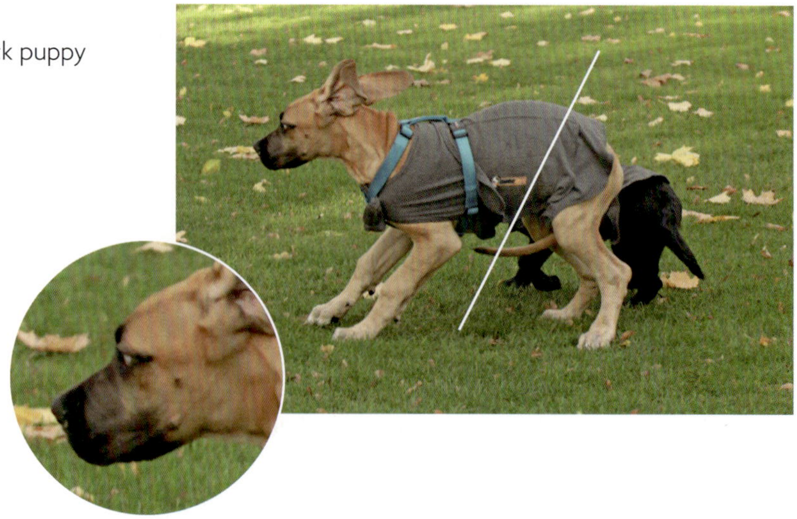

BALANCE POINT SHIFTS TO THE FRONT

Dogs usually shift their balance point towards the front when they are curious or on the offensive. If a dog is curious and cautious at the same time, the weight shift is marginal, with the hind legs 'left behind' the rest of the body, to allow for a quick get-away.

Cautious curiosity

Ella is a new member of Katja's family and has not yet met all the other animals. She is curious about Katja's cat, Molly, and cautious at the same time. She looks directly at her.

- Balance point towards the front

Additional clues indicating tension:
- Neck lengthened and head extended forward
- Ears held to the side
- Tail raised high
- Hind legs left behind the body

Stalking

- Balance point forward

Additional clues indicating tension:
- Head pushed forward and level with the back
- Tail lowered
- Body crouched

Playful stalking – with anticipation

- Balance point clearly forward

Additional clues indicating tension:
- Head slightly raised
- Tail held low and motionless
- Increased angulation of the elbows and hind end
- Starting to crouch

SWITCHING BALANCE POINT

SUBTLE SWITCH

In this series of photos, Inca, the Dobermann, switches her balance point while standing still to get a little closer to Chemsi, the Sloughi puppy. Inca is both curious and cautious. Chemsi's continuous eye contact tells Inca that she needs more space, and the larger dog responds by slowly shifting her balance point and leaning back. The movement is very subtle, as you can see by the lines drawn on the photos.

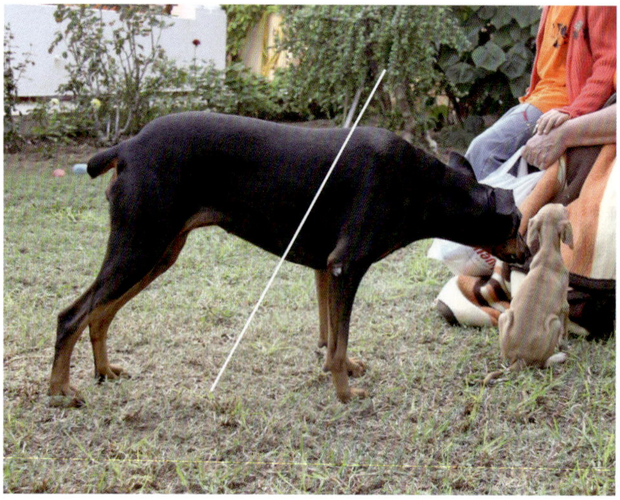

Inca, the Dobermann, has shifted her balance point forward so she can be closer to Chemsi. Most of her weight is now on her forehand. Inca is unsure about little Chemsi, and is struggling to figure her out.

Due to the strong eye contact from Chemsi, Inca slowly starts to shift her weight back. The weight of her body no longer rests on her front legs. Chemsi retains her eye contact.

While Chemsi maintains eye contact, Inca shifts her balance even further back and she also raises her head. Her hind end is starting to support more of her body weight. She uses the shift to increase distance from Chemsi, without moving her feet. The stub of her tail is lowered. She is quite excited, which you can see by the drool from her mouth.

OBVIOUS SWITCH

When two dogs meet you can tell by their shift of balance which dog wants to make contact and which would prefer to keep their distance.

When this male Labrador (*left*) meets Rani, the Golden Retriever, he leans slightly forward, and does not extend his neck as much as he could. However, it is clear that Rani thinks he is too close; she tries to create more distance by leaning back. Both dogs are excited, which you can see by the raised hairs on their backs. The facial expression of the Labrador is relaxed – he is blinking. Rani shows a lot of tension, her whiskers are raised, her eyes are rounded and her ears are pulled back and flattened.

The small, light-coloured dog, Tamika, is leaning forward with an outstretched head to make contact, whereas the larger dog is moving his balance point to the back. Both dogs are holding their ears in the flirting position – they are interested in each other.

The Golden Retriever (*left*) has shifted his balance point to the back while walking forward to show the small, light-coloured dog that his intentions are peaceful. The small dog is shifting his weight to the side and back.

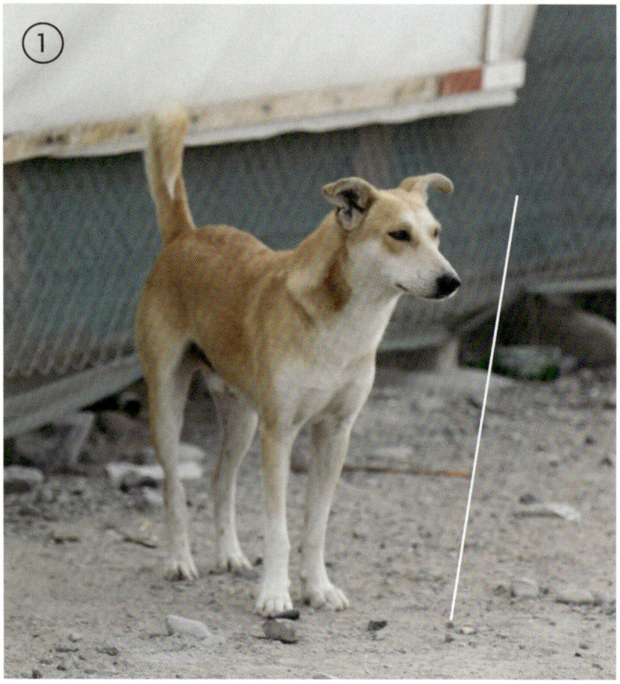

These photos were taken a split second apart. Looking closely at the first photo, you can see that the tan-coloured male is just about to move, and you can predict the direction he will choose. His balance point has already shifted forward.

Seconds later, the dog shifts his balance point towards the back in order to reduce speed. He has sighted another dog in front of him.

BALANCE POINT SHIFTS TO THE SIDE

There are some 'rude' dogs that invade the personal space of another dog by leaning sideways. Often, the other dog responds by leaning away in order to maintain the desired distance.

These two dogs are moving at speed, and their proximity may have been accidental. In a chance encounter of this kind, dogs will often lean to the side.

These dogs are comfortable with each other; they shift balance in harmony.

OBVIOUS BALANCE SHIFT TO THE SIDE

Often a shift of the balance point to the side is so subtle that you have to sense it, rather than see it. However, at other times, it is very clear.

The male Rhodesian Ridgeback (*left*) has a very rounded back and is leaning away. The female Weimaraner has been busy digging a hole. She shows from her facial expression that she is curious and interested in the Ridgeback, and is experiencing no tension.

The Ridgeback remains motionless, leaning away from the Weimaraner, and turning his head away, as she comes closer and makes direct eye contact. His ears are back and his tail is moving.

The tense and self-absorbed posture of the Ridgeback relaxes as the Weimaraner turns away from him. He turns his head even further away from her. He wants to stay neutral in this situation and is very careful not to provoke her by moving. Maybe he accidentally came too close when she was digging her hole, or perhaps she wanted to make contact, but he did not.

Rani, the Golden Retriever, is displaying a distinctive leaning away posture to indicate that she does not want to engage in a confrontation with Ella, who is barking at her. Rani does not want to provoke the much smaller dog by moving and so stays perfectly still until Ella leaves.

York, the adult Golden Retriever (*left*), is clearly leaning away from the puppy, Arwen, in order to create distance between them. Arwen is also being polite. She is leaning away from York to show that she respects the older dog's need for space. She also turns her head away.

Quivive (*left*) is creating more distance from Taylor, the big Pyrenean Mountain Dog, by leaning away. The angle of Taylor's pelvis is pointing towards Quivive, but the rest of his body is leaning away from her, indicating that he respects her need for distance. He is acting very politely, holding his head in a position that avoids eye contact, and blinking.

Arwen is torn between curiosity and caution when she meets Katja's Potbellied pig, Hotte, for the first time.

- Clearly leaning away to the side

Additional clues indicating tension:
- Neck very low
- Back rounded
- Tail dropped

As the pig turns away, Arwen's balance point also changes. She is in a more neutral position, no longer leaning to the side.

People also lean away to create more distance. Here a crawling baby spontaneously shifts her balance backwards when the dog, Tamika, turns her head and makes eye contact.

Interactions between dogs and small children should always be supervised by a responsible adult. Tamika is a very confident dog and will get up and leave when she feels she needs more space. It is very important to ensure a dog can always exit a situation. A dog who feels cornered, or restricted, is left with only one option – to go forward as a means of escape. Bear in mind, restrictions arise from confined space, lack of mobility (due to age, or injury), or a leash.

Often when dogs are interacting with each other, they swiftly change their point of balance as the situation unfolds. This is recorded in the following series of photos where two males experience an uncomfortable encounter:

- The light-coloured male is trying to mount the black-and-tan dog to show that he is the tough guy. He attempts to intimidate the black-and tan-dog by standing facing his side and placing his head on top of his back.
- The black-and-tan male is clearly indignant, but he stays calm, in spite of the other dog's rude behaviour. The light-coloured male keeps trying to use his body to make close contact.

The black-and-tan male, who shows his confidence with his erect posture, is telling the lighter coloured male that he needs to back off. He is a more mature, self-assured dog, who is trying to resolve the situation before it escalates.

WHAT IS SELF-ASSURANCE?

A self-assured dog is calm, confident and in control, and is able to rise above a given situation. He does not provoke, and uses de-escalating body signals to neutralise proceedings. He will defend himself when necessary, but, generally, he will use his confident posture to find peaceful solutions.

To the untrained eye, a mature, self-assured dog can appear to be fearful because he does not want to engage in a fight. When he decides that the situation cannot be resolved peacefully, he will usually turn and walk away. A guardian of this type of dog should support his behaviour and respect that the dog is capable of making very smart decisions.

OUR INFLUENCE

So far, we have primarily focused on shifting balance points in the context of interactions with other dogs. However, the proximity of people will also have a strong impact on this behaviour. Since physical, emotional and mental balance are very closely connected, we need to consider the effect that we have on our dogs.

EMOTIONAL TIPPING POINT

Hadiya, the Rough Collie, has issues with people who come too close to her, and she has bitten in the past. She would not normally allow anyone to come this close without attacking. The man, appears to be unaware of her history and stays perfectly centred over his own balance point.

The man is kneeling in front of Hadiya and his balance point rests on the knee he is using to support his upper body. His upper body is bent forward slightly, as he makes eye contact and talks to her.

Hadiya has shifted her balance point to the back. This shift of balance could mean that she is preparing to attack. She raises her paw, makes direct eye contact with the man, and keeps her tail close to her body. Her ears are flat against her head.

Here Hadiya has pulled back a little. She breaks the eye contact and does not attack. The man has turned his head slightly; he has not changed his overall position but does not invite further interaction.

Hadiya then turned and walked away. The man waited until she was at some distance, and then got up slowly and left. Shortly after this, Hadiya attacked another man, who just happened to be walking past. The man did not attempt to interact with her. It is not clear what triggered her behaviour, but it could have been a release of tension following the earlier interaction which she found highly stressful.

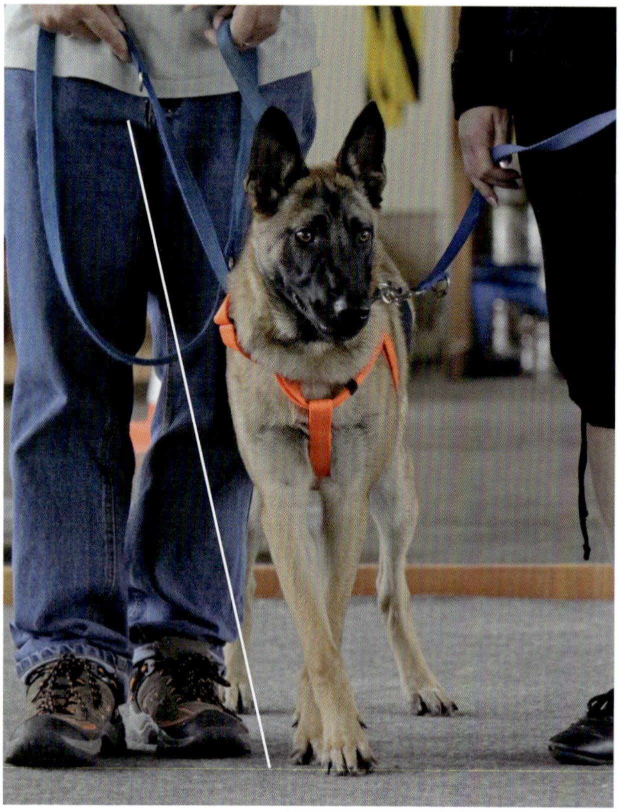

Loki, the Mallinois, has a tendency to be impulsive. He has developed a habit of leaning into his guardian's legs while standing still (*above left*), which means he is not finding his own point of balance. Working on the 'homing pigeon', a Tellington TTouch leading exercise, where the dog is balanced between two people, has helped him to find and control his own balance point. By learning to be better balanced physically, his emotional balance has greatly improved and, with it, his impulsive behaviour.

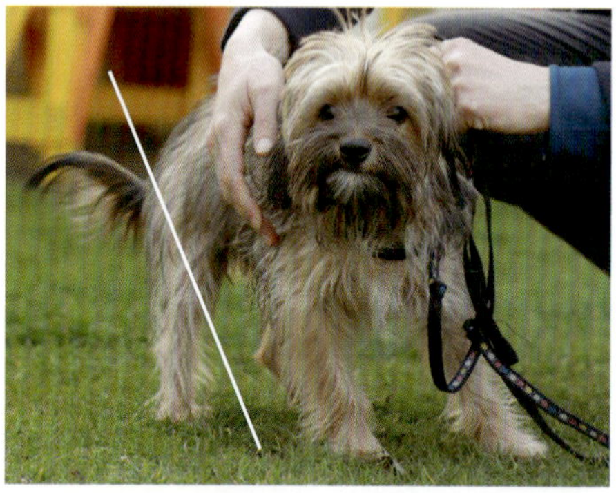

This little puppy appears to be very concerned. His guardian's knee, unintentionally, invades his space which worries him. In both photos you can see how he leans away from his guardian and, in the photo (*right*), he has a whale eye.

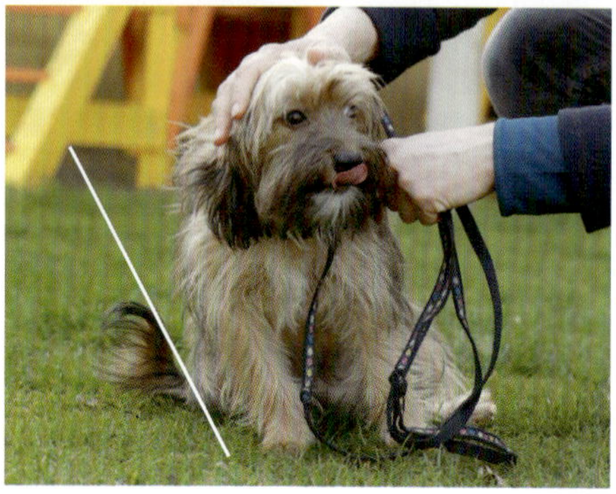

When his guardian reaches for him, over the top of his head, he licks his lips and nose (*left*). However, he has been able to create a little more distance from his guardian's knee, which has allowed him to bring his balance point closer to neutral.

PULLING ON THE LEASH

Leash pulling, and the tension that it creates in the body, are often the cause for so-called 'leash aggression'. The pressure on the collar limits the dog's ability to breathe and the brain receives less oxygen. Obviously, this is a bad starting point if a dog is trying to cope with a difficult situation.

This is the typical posture for a dog, who is being held on a tight leash. This stance can quickly become hard-wired for the dog.

Tension

- Balance point to the front

Additional clues indicating tension:

- Hind legs behind the body
- Front legs braced
- Tail raised
- Neck raised
- Head stretched forward

If a puppy is allowed to pull on the leash, this pattern of movement is stored in the nervous system and is very difficult to change later on. Obviously, the opposite applies to puppies who learn to walk on a loose leash. The Tellington TTouch leading exercises are an effective method of teaching good leash walking manners to the youngest of puppies.

If puppies are restrained on the lead, it is impossible for them to operate independently. These two puppies have already got too close to each other and cannot a find a way of creating distance.

When on leash, a puppy needs to receive the correct information, and be given enough time to respond to a change in pace or direction. Puppies should always be led with a well-fitting harness.

WORKING WITH DOGS ON LEASH

Most puppies love meeting and greeting other dogs when they are out on a walk. However, it is important to create sufficient distance so the puppy stays neutral and learns to be polite when interacting with other dogs. This distance will vary from dog to dog, and from situation to situation. Generally, adult dogs have a much greater need for space than puppies, and this needs to be respected.

A leash that is too short, and constantly interferes with the dog's natural speed while walking, can cause the dog to become an habitual puller. Dogs that have never learnt to walk on a leash, such as street dogs that are accustomed to choosing their own space and speed, have a natural tendency to pull on the leash.

Leash aggression is usually caused by muscle tension, which leads to emotional tension. Most dogs will not show this aggression when they are off leash. In addition to the high level of tension, lack of space, and the inability to use body language as a means of communication, are common triggers for leash aggression.

Continuous pulling on the leash can cause damage to the vertebrae and joints. When the leash is attached to a collar, the larynx, thyroid and trachea will also be negatively affected and can cause breathing problems.

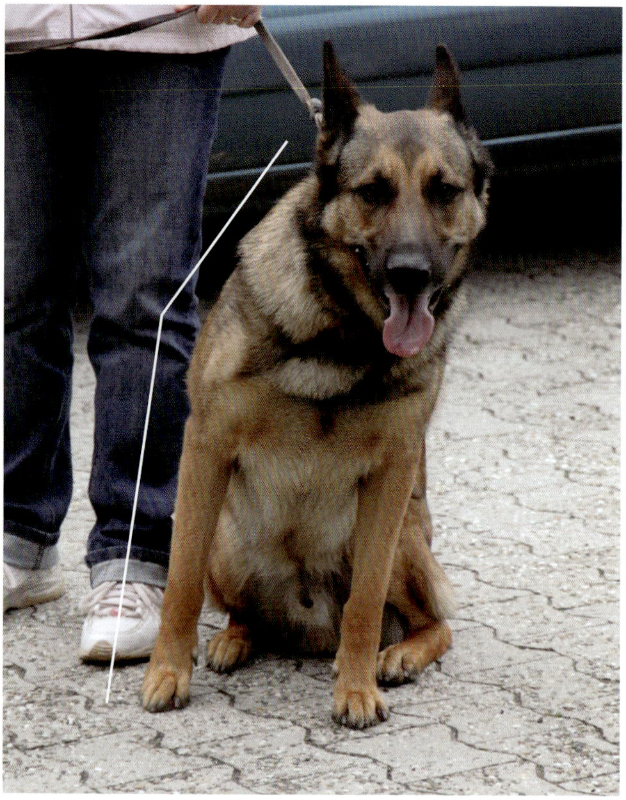

In these photos, both dogs are leaning to the side to try to stay in balance when there is a strong pull on the leash.

Here we have two Kangal Shepherd Dogs standing very close to each other. The male (*right*) wants to display his confidence. The other dog is very aware of his message, and shifts his balance point to the back. This is not an easy situation because both dogs are on leash and, therefore, they are not in balance. This greatly limits their body language and means they cannot resolve the situation themselves. They are in close proximity and are standing eye to eye with direct frontal eye contact.

It would be advisable to ask both dogs to turn away from each other so they break eye contact and can create more distance. When dogs are on leash it is very important to be mindful of the distance they need to feel safe. The need for space and distance is usually much greater for dogs on leash than when they are free. They need more space than we think.

This is a situation which occurs far too often. The two guardians are engaged in a conversation and have forgotten about their dogs.

This is problematical for the two dogs as there is very little space between them. The German Shepherd, pictured right, is keen to meet and greet whereas the other Shepherd is more cautious. This is not helped by his guardian maintaining a tight hold on his leash, which is adding to his sense of tension.

The Shepherd (*left*), would like to leave, but the leash makes this impossible. He is clearly uncomfortable, which we can see by his pulled-back ears and long mouth. In the second photo, he shifts his body weight as far back as possible. His topline is rounded and his tail is very tight. This type of situation can escalate quickly.

In this scenario, the two leashes have got tangled and the dogs can no longer avoid each other and preserve their space. It is important to remedy this situation quickly.

The Bulldog (*left*), is hanging on his leash; his front feet are not even making contact with the ground. This can cause too much pressure on the trachea and deprive the dog of oxygen. Lack of oxygen will mean the dog is unable to think, or act, properly.

The Border Collie (*left*), has thrown his point of balance so far to the side that he no longer has control over his own body. If his guardian suddenly loosened the leash, he would probably fall over.

THE OPPOSITION REFLEX

It takes two to pull on a leash. The opposition reflex is something all mammals possess, and it is responsible for keeping the body in balance.

Therefore, we, too, have an opposition reflex and we react accordingly when our dog pulls on the leash. We lean against the pressure and answer by applying the same amount of pull, because force creates force. To hold the leash in a relaxed manner, we must not trigger our own opposition reflex in response to the pull from the other end of the leash. Tellington TTouch shows ways to leash-walk the dog without triggering his opposition reflex, and if you did so unvoluntarily, it shows you ways to resolve the situation.

Here the handler's pull on the leash is so strong it causes the dog to shift his balance and exert as much strength as his handler.

This herding dog cross is very much out of balance – and so is his handler.

Focusing on the herding dog, on leash, his body weight shifted to the back ...

... ready to leap...

... and lunge forward.

GAINING CONTROL

Did you know that improving physical balance will automatically improve emotional and mental balance?

There are many ways to teach a dog to have more awareness of his body and, therefore, give him the tools to gain control over his centre of balance. This will, in turn, improve his emotional confidence and his sense of security. The seesaw is an effective means of teaching a dog about his balance point. Work on the seesaw should be taken slowly, and we recommend having the equipment on its lowest height.

Quivive had a bad experience with a seesaw when she was a puppy and, as a result, she developed a fear of unstable or moving surfaces. She was reluctant to step on to the seesaw and really did not want to be near its tipping point. When the board moved downwards, she lost her physical and emotional balance. Over time, she learnt to negotiate the seesaw, and gained confidence in her ability to shift her balance point. In this way, she discovered that she could control the tipping point of the obstacle.

Quivive is very tense and tries to enlarge the tread area on the seesaw by spreading her hind legs far apart in order to gain more stability.

By breaking the exercise into small steps, Quivive has learnt to walk confidently to the tipping point, stop, and then shift her weight to the back so the seesaw could not tip.

From this position, she can shift her weight forward by raising her hind end. Since she is a very light dog, she has to use all of her body weight to get the seesaw to tip. In fact, she almost lies down. The board starts to tip while she is crouching low and is in good control of her balance. During practice, she wears a Tellington body wrap to support her body awareness.

By conquering an obstacle that caused a lot of concern, Quivive gained both physical and emotional confidence. She became braver and faced new situations with a more confident attitude. Her trust and connection to Gabi was also strengthened.

6. LYING DOWN

The positions in which a dog lies down to rest, or sleep, gives us a lot of information about how well and how secure he is feeling. When he is lying down, and is truly relaxed, he is experiencing a high degree of wellbeing and feels completely safe.

RELAXED

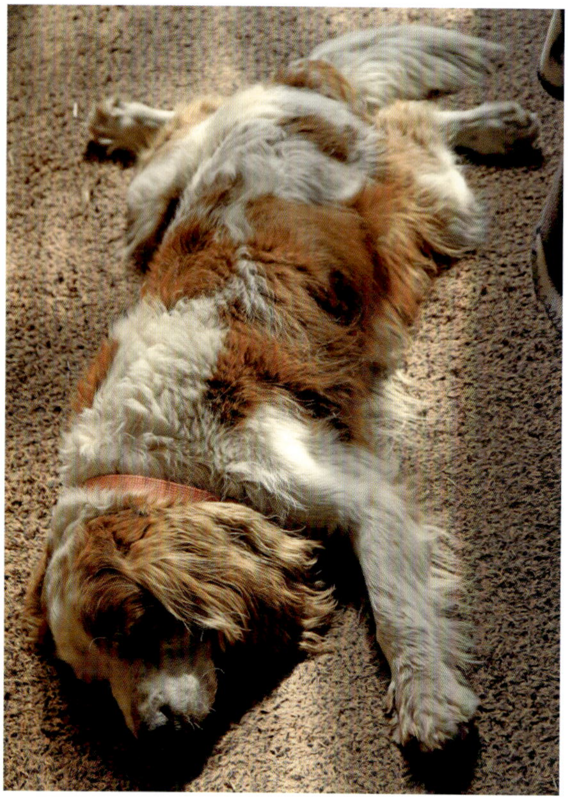

Every dog has his own way of feeling comfortable. Some dogs, like Edward, the Cavalier King Charles Spaniel, like to lie with their legs spread far apart (frog legs). He is very relaxed.

Dhanyi likes to rest with her hind legs stretched far behind, her as she is showing us on this park bench.

Quivive was very relaxed in the midst of a number of other dogs who attended a Tellington TTouch® course. Here she is being the demo model for the Turtle-TTouch that is being applied to her front leg.

Signs of relaxation

- Front legs, neck and head loosely on the floor
- No physical tension
- Eyes closed
- Ears tipped to the back

Many dogs like to be covered with a blanket. Some just find it comfortable, for some it serves as a security blanket, and others enjoy the warmth.

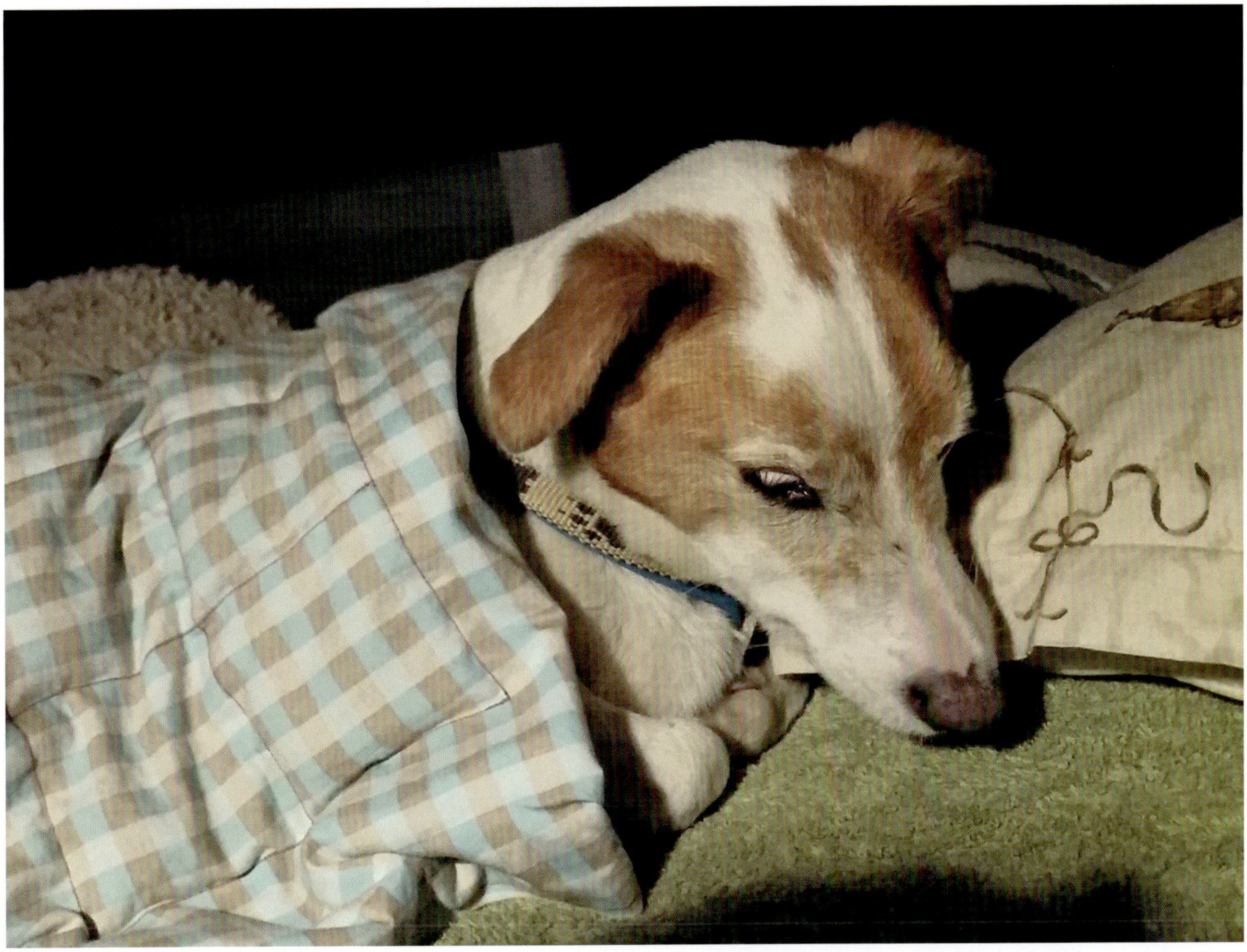

Michel, the terrier, has been rehomed from a shelter. It is likely that he had distemper; he shows some distortion of perception and tends to be very restless. He refused to allow his hind end to be touched. However, when he was covered with a heavy blanket he could relax more easily. His guardian was familiar with the calming effect of weighted blankets through her work with people, so she sewed rice-filled pockets into Michel's blanket.

LYING IN CLOSE CONTACT WITH PEOPLE

Many dogs enjoy close contact with the people they know, and sometimes they will even relax with strangers.

It may not look comfortable but Ella is small enough to enjoy close contact with Katja in any position.

Atrejo, the Irish Wolfhound, is just a little too big to climb on to his guardian's lap.

During her first camping trip, Quivive, the Pyrenean Sheepdog, enjoys the safety of a familiar lap.

It is important to bear in mind that not every dog enjoys close contact with a person when he is resting. Regardless of our own feelings, we should respect the fact that some dogs need space, and time alone.

LYING IN CLOSE CONTACT WITH OTHER DOGS

For puppies it is natural to lie and sleep in very close contact with their littermates. This behaviour may also be seen in adult dogs who live together.

Maxima (*left*) and Carlotta live together. Carlotta is very insecure and lying close to Maxima gives her a lot of emotional support.

Tamika and Rani are part of Katja's large family of dogs. Here they lie together contentedly, sharing the limited space available on the sofa.

A dog will roll on to his back when he feels safe in his environment. The dog needs to feel very relaxed as it is impossible to react quickly – either taking flight or going into defence mode – when sleeping in this position. Dogs who are sick, or elderly, are usually too stiff to adopt this position, even when they are feeling secure.

Ischka enjoys lying with the cat, Heller.

GROUP SHOTS

Harmony

Here the dogs are in close contact, and you can see from their expressions that they are perfectly happy with the situation. This type of close contact is only possible with dogs who know each other, and are relaxed in each other's company. Due to their daily Tellington TTouches® Katja's dogs are relaxed and they get along with each other splendidly.

Discord

The day before this photo was taken, the two entire Australian Shepherds (*centre*), got involved in a serious fight which also upset the other dogs. When they were positioned for a group photo, the following day, the two males show their discomfort by leaning away from each other. The two females, on either side of them, reflect the tension in the atmosphere by leaning away; the female (*far right*) is also blinking. Neither want to become involved in the brothers' issues. Their guardian decided to have both males neutered, which restored harmony between the brothers.

Physical contact of a different type: Inca, the female Dobermann, shows little concern even though Chemsi, the female Sloughi, is sitting on her. You will meet Inca and Chemsi many times in this book. Chemsi joined the family, which included Inca, when she was a puppy. Initially Inca was very insecure. In contrast Chemsi was a very confident puppy, and had no problem finding her place in the family.

TENSION

When tension in the body increases, a dog's posture will change and different muscles will become tight. This can occur for many different reasons. Some dogs (and this can be breed specific) have higher tension in their muscle tone, which occurs naturally. This can be seen in an inability to rest all four limbs on the ground when lying on their side; they may also struggle to release the tension in their neck muscles. This chronic physical tension has a knock-on effect and raises the dog's emotional tension.

Partially relaxed: Bonnie is starting to relax after a Tellington TTouch session. Her left hind leg is pretty close to the ground. She is still holding her left elbow by turning it to the outside, which prevents her left front leg from making contact with the ground.

This Vietnamese street dog remains lying down when he sees the photographer, but his face shows slight tension.

Slight tension – aware

- Head resting

Additional clues indicating slight tension:

- Ears raised at the base
- Eyes open
- Eyebrows raised lightly
- Whale eye
- Looking directly at the photographer

Chocolat is lying beside her guardian in a half-raised position, watching everything that is going on with great intensity. She has raised her upper body ready to spring to her feet, if required.

Tense – guarded

- Pelvis tipped
- Lying on her elbows
- Upper body raised
- Neck raised high
- Ears raised at the base
- Eyes focused forward
- Mouth closed

Some dogs need to be in control and have a strong urge to interfere if another dog, or another person, comes too close to them or their people. The high degree of tension shown in Chocolat's posture (above) reflects this attitude. In this scenario, his guardian should encourage him to switch focus and change posture, by talking to him, or making gentle contact with him.

PHYSICAL LIMITATIONS

If a dog suddenly changes his normal lying position, it could be an indication of physical pain or illness.

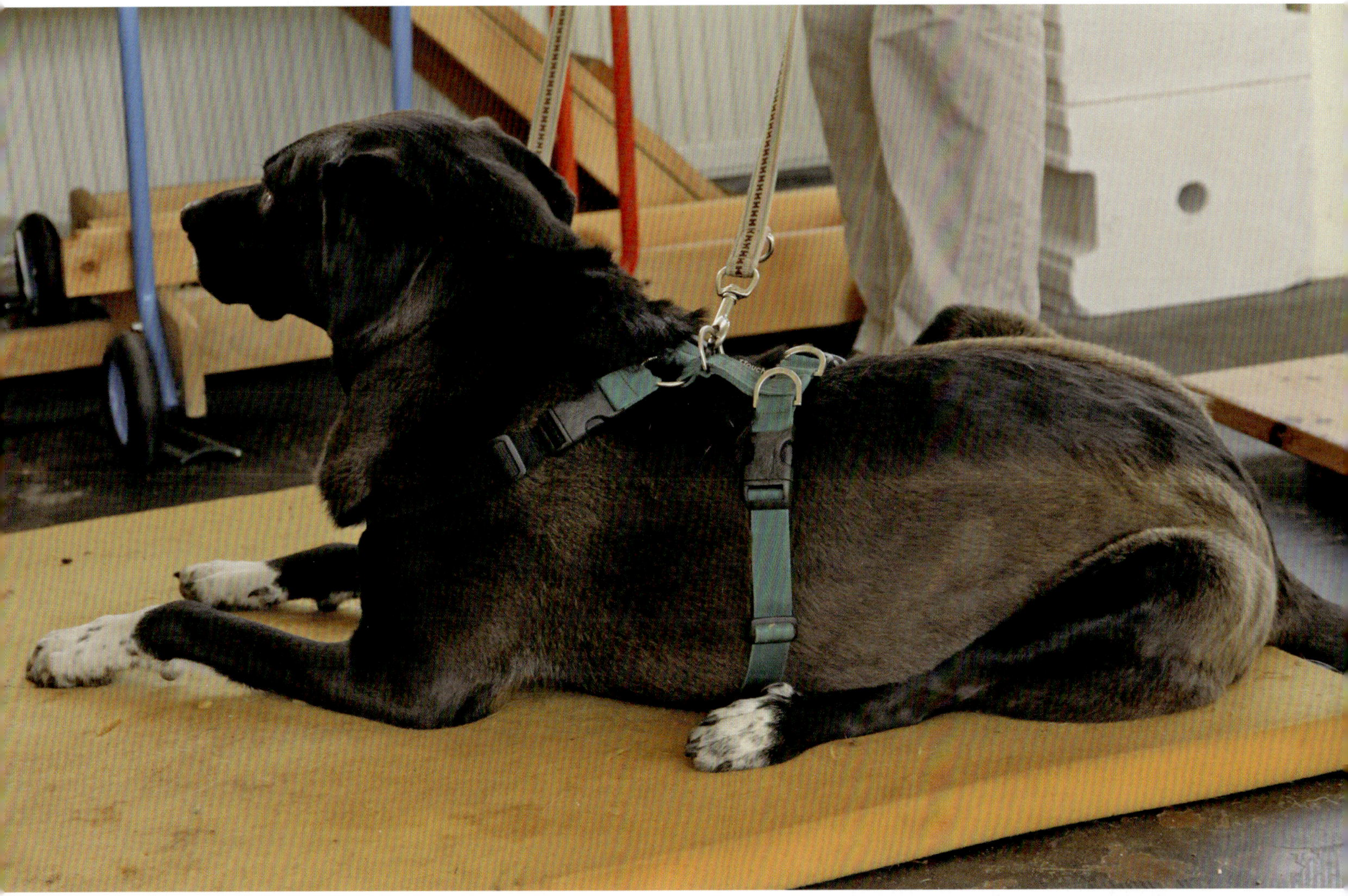

Bonnie is making himself comfortable on a foam pad which is actually intended for dogs to learn to walk on different surfaces. Bonnie fractured his leg when he was a puppy, which required surgery. This resulted in one leg being shorter, so he is no longer able to place his left leg under his body when lying down. Older dogs, or those with a physical or anatomical issue, appreciate access to a soft surface that adjusts to their body and prevents pressure points. Nowadays there are a lot of high quality orthopaedic pads on the market.

Lying position after a leg fracture resulting in a shortened right leg

- Left hind leg turned towards the outside
- Weight on the point of the elbow
- Front legs not completely resting on the ground

Pancreatitis can cause severe pain that also moves into the back. Quivive is not showing the other typical symptoms, such as vomiting and diarrhoea, and her pancreas specific lab work stayed in the normal range for a long time. Therefore she was treated mainly for her back, which was very sensitive to any pressure. When this photo was taken, she was in a lot of pain.

Tensed lying position with hind legs pulled close to the body due to severe pain caused by pancreatitis

- Back and neck severely rounded
- Head not resting
- Hind legs pulled close to the body
- Right elbow braced away from the body
- Front leg not making contact with the ground
- Front paw braced

As he aged, Tiber suffered from spondylosis, which is a degenerative ailment of the spine accompanied by the development of painful bone spurs.

Left: When the pain was severe he rested in the sphinx position to eliminate some of the discomfort.

Right: When he was feeling a little bit better he was able to lie more on his side, but he wasn't strong enough to lie completely flat.

Some dogs adopt unusual positions when sitting or lying down, as illustrated by Sandy, a Podenco mix. According to her guardian, she usually sits or lies like this. Her harness is too large and has slid down over her shoulders.

If you think your dog's resting posture looks tense and uncomfortable, he could be feeling unwell. Many sighthounds feel the cold and will relax with the added warmth of a blanket. Many dogs dislike hard surfaces. It is important to keep an eye on your dog's resting position to ensure he is comfortable and feeling well.

Harnesses for dogs are like saddles for horses. They have to fit the individual dog and need to be adjusted to specific needs. The correct fit and design of the harness is important so that you can guide your dog with a loose leash without jeopardising your own and your dog's wellbeing.

PHYSIOLOGICAL FACTORS

by Dipl. Biologist and Tellington TTouch® Instructor Karin-Petra Freiling

The homeostasis, or the balance, of the physiological functions of the body have a profound effect on a dog's behaviour and wellbeing. The stress that a dog may experience in his everyday life can lead to exhaustion, which can be a stepping stone to illness.

When a dog is stressed, the limbic system of the brain takes control. The cerebrum and, therefore, the ability to think and evaluate situations rationally, is turned off. The liver works at full capacity to supply the body with everything it needs to fight or flee.

Cortisol and adrenaline flood the system. Adrenaline is guided directly into the bloodstream. It takes about six hours until the cortisol has reached its maximum level, and it can take up to six days for the body to dissolve the excess and bring the cortisol levels back to normal. During that time it simmers inside the dog, affecting his behaviour patterns. So, if a dog experiences a stressful event, it is important to encourage relaxation by using the Tellington TTouch® Method.

Providing high-quality food, which includes the substances that the body uses up at times of stress, is very important. Glucose, in the form of fruits and high-grade carbohydrates, will help the over-worked liver to function efficiently, as well as benefiting the rest of the body.

As the photographs in this book illustrate, the stress a dog experiences is readily identified in his body language and behaviour. If you are able to detect this sometimes subtle form of communication, you will be able to help your dog regain his physiological balance and restore his sense of wellbeing.

The Tellington-Method, with all its facets, is ideally suited to bringing a dog's body back into balance. The gentle Tellington TTouches® create a surge of oxytocin, which is the biological messenger for the trust hormone. Oxytocin is best known as the hormone that strengthens the bond between a mother and her child after birth.

DECREASING EMOTIONAL TENSION

If a dog is in an unfamiliar situation, it can be very hard for him to lie down and relax. He will usually lie with his legs underneath his body instead of lying flat on his side. Alternatively, he may choose to lie with his upper body raised because he is too tense to relax his shoulders and neck, and lie with his head on the ground.

In these situations, we have found that the Tellington TTouch® Method can encourage dogs to be more relaxed and settled.

Nala and her guardian attended their first HUNDGESTÜTZT clinic in Berlin, Germany where the Tellington TTouch® Method is also taught. At the beginning of the course Nala was tense when she was close to the other dogs. In the following sequence, she is receiving some TTouch on her front legs. Nala is the type of dog that has so much natural muscle tension, she is unable to rest her legs completely when she is lying down.

Severe tension
- Head raised
- Top of the head wrinkled
- Ears raised at the base and held to the side
- Looking straight ahead with strong focus
- Eyebrows raised and pulled towards each other
- Mouth closed
- Front legs not quite touching the ground

Slight tension
- Head lowered slightly
- Top of the head wrinkled
- Ears raised at the bottom and held to the side
- Eyes not as focused anymore, looking down
- Mouth closed
- Front legs not really touching the ground
- Body weight towards the front

The Ear TTouches changed two things for Nala: She was able to lower her head and no longer needed to stare straight ahead, and she was able to switch her focus and look down.

Nala started to relax a lot more when her guardian used TTouches on her head, neck and shoulders.

Increased relaxation

- Head is resting on the ground
- Neck is still slightly raised
- Top of the head smoother
- Very few wrinkles
- Ears hanging more
- Look of the eye unchanged
- Mouth closed
- Left front leg is not making full contact with the ground

Nala is getting more relaxed and she is even able to close her eyes.

Her guardian is using Ear TTouch and you can see how the TTouches have relaxed her neck muscles to a point where she is able to let her head lie down between her front feet.

Optimum relaxation

- Head resting on the ground
- Neck released and lowered
- Top of the head noticeably smoother
- Base of the ears no longer raised
- Eyes closed
- Position of the front legs unchanged

Due to her skin type, Nala shows wrinkles on her face and on top of her head when she is experiencing minimal tension. It is important to be mindful of this characteristic otherwise the presence of wrinkles could be be misinterpreted.

Formerly, Nala was known for her high-energy outbursts but, following her TTouch training, she has grown in confidence and is able to respond calmly when faced with difficult situations.

TTOUCHES FROM LINDA TELLINGTON-JONES

In the following sequence of photos, Quivive is receiving treatment from Linda Tellington-Jones, in person. A dog can experience a lot of muscle tension, leading to to physical limitation, an uneven gait, and pain, even though he is not suffering from a specific injury. This will not only affect the dog's physical state, but also his emotional wellbeing. The Tellington TTouch® Method is a perfect way of supporting veterinary, naturopathic or physio treatment the dog may be receiving.

Quivive often finds it hard to relax her left hind leg. She holds her leg up high, and can release it only to a certain point, before jerking it up again. When she is in an unfamiliar situation, she finds it especially difficult to release the tension in her muscles. At the beginning of the TTouch session, she holds her leg very high.

While Linda is TTouching Quivive, the tension in the muscles of her hind leg releases. She rests her leg in Linda's hand and Linda gently guides the leg on to the blanket.

Quivive is deeply relaxed after her TTouch session and is lying on the table with her eyes almost closed.

7. THE TAIL

From an anatomical perspective, the tail is a boney extension of the spine.

There is a wide variety of tail types:

- Hanging – seen in Golden Retrievers and German Shepherds
- Raised – seen in Dalmatians
- Curled – seen in Pugs, Alaskan Malamutes and Spitz breeds
- Natural bob – seen in Australian Shepherds and Bulldogs

TAIL TYPES

The most common tail type, which is seen in many breeds and mixed breeds, is the relaxed hanging tail with a slight curve at the tip. This type of tail can change dramatically when a dog is tense.

A sword-shaped tail is commonly seen.

There are some breeds, such as the Pug (*left*) and the Pomeranian (*right*) that have short tails, which are carried tightly curled over the back.

Some breeds have naturally short tails, such as the Fox Terrier (*left*). Others are born with a bob tail of different lengths, such as the Mini-American Shepherd (*right*).

TAILS AND BALANCE

Tails play an important role in balancing the body, and are also used as a means of communicating emotions. Thankfully tail docking, which was common practice in a number of breeds, is outlawed in most parts of the world but it still exists in some countries to the detriment of the breeds concerned.

Quivive has no physical limitations but she still uses her tail to help her to balance as she negotiates a frozen puddle.

Here a Yorkshire Terrier uses his tail to balance on the log while interacting with another dog.

Luke has suffered a femoral compound fracture, with nerve damage. He can no longer use his right hind leg but is able to maintain reasonable mobility by holding his tail to the right to aid his balance. These photos, taken within a one second time span, show how much Luke relies on his tail.

COMMUNICATING EMOTIONS

The position of the tail, and its movement or lack of movement, gives us a lot of information about a dog's mood at any given moment. Obviously, this needs to be viewed in conjunction with his posture and associated body language, and the situation in which he finds himself.

A wagging tail, for example, does not simply mean a dog is happy. He is excited, and this can be the predictor of many different outcomes.

The height at which he carries his tail, the speed at which he wags it (frequency), and how far he swings it from side to side (amplitude), give us more information which will indicate whether he is feeling friendly, fearful or aggressive. If a dog moves his tail up and down, instead or side to side, it is generally the result of tension or a physical anomaly.

MOOD BAROMETER

Slight tension: The dog wags his tail and swings it softly with larger amplitude. The tail is usually carried about level with the back or a little bit higher or lower.

Increased to very high tension: The dog wags his tail faster and with shorter amplitude. The tail is also carried higher. When a confrontation is looming, and there is a lot of tension, tail movement is minimal. The dog will hold his tail so that it is high and still, and very stiff. Sometimes he will move the tip of the tail very rapidly.

Happy: The dog usually holds his tail very high and moves it fast. Some dogs even move their tail in circles when they are really excited. A happy dog may also keep his tail low and wag it quickly, similar to the tail carriage when he is feeling insecure.

These three dogs are focusing on the same thing, but they are all carrying their tails differently.

RESPONDING TO DIFFERENT SITUATIONS

Dogs carry their tails at different heights in different situations not just because of their conformation and breed, but also due to their individuality. A dog's personality and confidence will determine if he carries his tail casually, or whether he uses it to appear provocative and intimidating.

Rani, the Golden Retriever (*right*) does not change her body posture when she meets the male Labrador. She does not crowd him and her softly swinging tail signals her friendly intentions.

Relaxed tail: Friendly-interested contact

Rani (right):

- Tail level with the back, swinging softly
- Neck raised
- Ears hanging loosely to the side
- Balance point neutral

Rani, the Golden Retriever, is looking over top of the Dachshund. The hair on her back is raised, she is slightly tense but would like to stay neutral.

Tense tail: Maintaining distance

Rani (right)

- Tail raised high
- Movement restricted to the upper third

The male (*left*) shows his intent to provoke and intimidate through his posture and direct eye contact. The guardian of the dog (pictured right) stepped in and took her dog away in an attempt to de-escalate the situation.

Very tense tail: Confrontation

Male (left)

- Tail raised high and curled over the back
- No movement to the tail

Rani has seen, or caught the scent of ducks. Her upright tail shows her extreme excitement.

Extreme tension in the tail: Intending to hunt
- Tail vertical, no movement

Additional clues indicating tension:
- Ears pointing forward
- Nostrils wide open

This black female lives in a shelter and here she is eagerly awaiting the arrival of a kennel worker. During her entire stay, she remained relaxed and confident. She did not let the behaviour of any of the other dogs influence her, even when their reactions were intense.

Relaxed at the fence
- Tail hanging low
- Balance point over the middle of her body
- Neck and head raised
- Ears carried loose, pointing to the side

Now the black female has been joined by two of her kennelmates.

Alert at the fence
- Tail carried level with back and moving
- Point of balance slightly forward
- Neck and head raised
- Ears carried loosely to the side

When approached by the photographer, the black female carried her tail low and made herself small. She used to be a street dog and always approaches strangers with caution. After a little while she became more trusting and sought out closer contact.

Cautious approach
- Tail carried low and moving
- Hunched posture (crouching)
- Neck level with the back
- Head stretched forward
- Ears held towards the back
- Hairs raised in the shoulder area

TAIL MOVEMENT

Frequency and amplitude

Neither the height of the tail, nor the speed of the wag, gives you all the information you need to evaluate a dog's emotional state. The entire situation needs to be considered.

This dog is enjoying a trick training session, and is eagerly awaiting his next cue.

High tail wagging – excited anticipation

- High tail carriage with high frequency

Additional clues for excitement:

- Neck raised high
- Ears raised at the base and held towards the back
- Eyes slightly rounded
- Direct eye contact

Arwen, the Schafpudel, is very excited when children she knows come to visit. This is very typical of her behaviour when she wants to belong.

Low tail wagging – happy

- Low tail carriage with lots of movement and great amplitude

Additional clues for excitement:

- Rounded back
- Crouching

Here Rani, the Golden Retriever is meeting a raccoon. Her tail is very low and moves only at the tip, at high frequency. Taking on board her posture and her ear carriage, we deduce that she is curious even though she is being cautious in her approach.

Low tail wagging – cautious curiosity

- Tail held low and close to the body with lots of movement

Additional clues indicating tension:

- Mouth closed
- Ears flattened against the head
- Point of balance towards the back
- Direct eye contact

This dog is being lured to walk between her guardian's legs. The presence of her guardian, leaning over her, makes her indecisive and insecure – but she would really like the treat. In the end, she is not brave enough and backs away. It is worth noting that you could make this exercise easier for the dog by dropping the treat, so you don't have to lean over so much.

Low tail wagging – insecure, paired with expectation

- Tail carried low and moving

Additional clues indicating tension:

- Back very round
- Point of balance shifted towards the back

This Sloughi puppy has just seen a stranger and he is feeling insecure. In this photo you can clearly see the rapid wagging of his tail as well as the large sweeping motion of each wag as it goes from side to side. He is cautious, but he is also showing curiosity as, generally, he likes interacting with people. A little while after these photos were taken, he sought contact with the stranger.

Low wagging with wide sweeping movement – caution paired with curiosity

- Tail slightly above the level of the back
- Sideways wagging with great amplitude
- Ears held towards the back
- Point of balance towards the back
- Mouth closed
- Eyes slightly rounded
- Left front paw raised

Wagging the tip of the tail

This Husky cross from Tunisia has seen something that has caught his interest. His body and head are raised. His tail is carried high; only the tip of his tail is moving.

He has spotted a family of wild dogs and the puppies are barking at him.

Dogs such as Bulldogs, that have very short bob tails, or corkscrew tails, are still able to move them from side to side. However, this subtle movement is not easily recognised by other dogs.

The white Bulldog (*left*) has lowered his tail; the piebald Bulldog (*right*) has raised his tail.

ASYMMETRIC TAIL MOVEMENTS

THE BRAIN AND TAIL WAGGING

In a scientific study in 2007, it was discovered that dogs wag their tails more to the left when they see a dog they don't know, and more to the right when they see a person they love. This is connected with activating the right side of the brain when wagging the tail more to the left, and activating the left side of the brain when wagging to the right. However, these findings are only accurate when the tail is equally flexible on both sides.

This is illustrated in the photos above. Tamika loves children and, looking from her perspective, she is wagging her tail more to the right, as described in the study.

Source: Asymmetric tail-wagging responses by dogs to different emotive stimuli Quaranta A, Siniscalchi M, Vallortigara G., 20.03.2007, Seeing Left- or Right-Asymmetric Tail Wagging Produces Different Emotional Responses in Dogs Marcello Siniscalchi, Rita Lusito, Giorgio Vallortigara, Angelo Quaranta, Published Online: October 31, 2013.

Rani, the Golden Retriever, sees an unfamiliar Dachshund in the distance. Her first reaction is to stop. She holds her tail high, barely moves it, and carries it a bit to the left. The Dachshund stops his approach and starts sniffing the side of the trail.

Rani lowers her tail and starts to wag it with a clear amplitude to the left.

TAIL CARRIAGE FOR COMMUNICATION

In combination with the rest of the body, the tail expresses a dog's willingness to communicate in the moment, and is, therefore, a predictor of how an interaction is likely play out.

Tails carried high indicate a certain amount of tension or excitement. The frequency and amplitude of the movement are individual and dependent on the situation.

A soft wag, with a low carried tail, indicates a friendly mood. This form of communication is often used when there is polite interest between the dogs who are meeting, or when a dog is meeting a person. The low wag, with a higher frequency, is shown when a dog approaches with caution. It is also used in combination with a play position when the dog is still hesitant about joining the game. *See Bowing, page 358.*

A tucked tail is evidence of a dog whose emotions may range from mildly insecure to absolutely terrified.

The leader of the Sloughi family (the female in the middle) has just returned from an outing and the other members of her family greet her. The white livestock guardian dog mix makes up to the older matriarch to show that he belongs to the family. *See Young dog advances, page 23.*

Gentle wagging as a respectful greeting

White dog:

- Tail held level with the back, swinging
- Eyes almost closed (blinking)
- Ears held towards the back
- Licking the corner of the mouth
- Point of balance towards the back
- Raising front paw

The darker male is interested in Rani, the Golden Retriever. She would like to interact, but is respectful. To encourage her, the male turns his head away and breaks eye contact. He keeps his head in a lowered position, and wags the tip of his tail.

Rani comes closer to him with a low swinging tail.

Tail carried low, swinging wide, for a cautious meeting

Rani (standing on the left)

- Tail carried below her back and moving

Additional clues indicating tension:

- Head held low
- Back slightly rounded
- Hunched posture (crouching)

These three dogs make direct contact with each other. Slight tension has arisen due to their mutual interest.

Slight tail movements when meeting neutrally

The three dogs:

- Toplines straight to lightly sway backs
- Heads held at the same level as back
- The ears of the Rhodesian Ridgeback (mid-group) are raised at the base and held to the side

This photo was taken in the United Arab Emirates and shows a street dog approaching a construction worker. The building site is close to the dog's den, and workers pass by every day. Sometimes they throw food scraps to the dogs, but often they are more menacing.

Tail low indicating indecision

- Tail lower than the back and wagging in large sweeping movements
- Low neck and head carriage
- Ears flat against the head
- Long mouth
- Back rounded
- Hunched posture (crouching)

Puppy, Tristan, meets Katja's ducks for the first time. He is curious, but undecided. He doesn't approach the ducks, but he doesn't move away. He is holding his tail high and only the tip is moving.

The ducks are used to meeting dogs and are equally curious. The mother of the ducklings (she can be recognised by the red around her eyes) tries to keep Tristan at a distance to protect her babies.

Rani keeps her composure as the young dog approaches her with boundless enthusiasm. Her posture clearly tells us that she sees the approach as impolite. *See Slowing down, page 299.*

High raised tail, no movement, due to ill-mannered approach

Rani (standing on the right)

- High tail carriage with little movement
- Neck raised
- Ears flattened and pulled back

The following sequence of photos show three dogs meeting at a lake. They are interested in each other, but there is a degree of tension which can be identified by posture and tail carriage.

The Rhodesian Ridgeback, who is standing on the right, keeps his tail straight up when he meets the black dog for the first time, and keeps his distance.

The light brown dog, standing on the left, raises his tail so it is almost vertical. The tail seems to be very stiff with little movement. His hair is raised in the shoulder area and his point of balance is forward. He is very excited and his focus is directed towards the black dog.

The black dog presents a moderately raised tail with little movement. His point of balance is towards the back, the joints in his hind legs are relatively bent. His mouth is long and he is a little insecure about the confrontational approach of the light brown dog.

FROM NEUTRAL TO SLIGHTLY TENSE

In this meeting, the tails are used to communicate.

Jack, the spotted dog, usually carries his tail very high when meeting a dog for the first time. However, in this situation he wants to reassure the female, Diana, who is somewhat shy. He shows his intentions are friendly by lowering his tail and slowing it down. His head and neck are also lowered and he shifts his body weight away from her.

The first contact is successful so Jack raises his tail, returning it to its natural position. He is not being pushy and keeps his distance from Diana who is often insecure around bold dogs.

Both dogs are on leash, which makes the initial meeting more difficult, because their space is determined by the people, and not the dogs themselves. The leash greatly limits a dog's ability to move and, thus, to communicate as he would wish.

FROM NEUTRAL TO VERY INSECURE TO SELF-CONFIDENT

The following pictures capture the changes of emotions in the female Golden Retriever, Rani, while meeting unfamiliar dogs. Her tail tells the tale.

Katja's family of dogs meet a large brown male (left), who approaches the female Golden Retriever, Rani, with his tail held high. The male is interested, but he is being cautious. Rani is very excited and carries her wagging tail level with her back. Her point of balance is slightly towards the back and the hairs on her back are raised. Tamika, the light-coloured smaller dog on the right, watches the interaction at a safe distance with her tail held high.

When the male makes direct contact, Rani pulls her tail between her legs and underneath her belly.

When the black female, Taschka, approaches with her tail held high, Rani's tail carriage changes and she now lifts her tail as well. Her hair is still raised. The male backs off a little; his balance point is clearly shifting back. He keeps direct eye contact with Rani and his tail has not changed, it is still high with very little movement.

TUCKED TAILS

A dog who is struggling emotionally will often carry his tail tucked close to his body and may clamp it so tight that it goes between his hind legs and touches his belly. When in this position, the tail will scarcely move. There are a number of reasons why a dog might do this:

Tucked tail while being bullied

The Stafford type, Helge (*left in the photos*), meets a small, female Pug in the street and shows his interest. He holds his tail very upright, and only the tip is moving. He is very pushy towards the much smaller female. The Pug is indecisive; her tail is usually curled up over her back, but she has straightened it and carries it low and tucked, almost under her belly. Her back is round and there is a strong angulation in her hind legs. She is trying to make herself small. The Pug is tolerating Helge's basic advances, but she is becoming increasingly insecure.

FROM HANGING TO TUCKED TAIL DUE TO INCREASING INSECURITY

This Hovawart is at a conformation show. He is not very comfortable in the ring, which he shows through his tail carriage and posture.

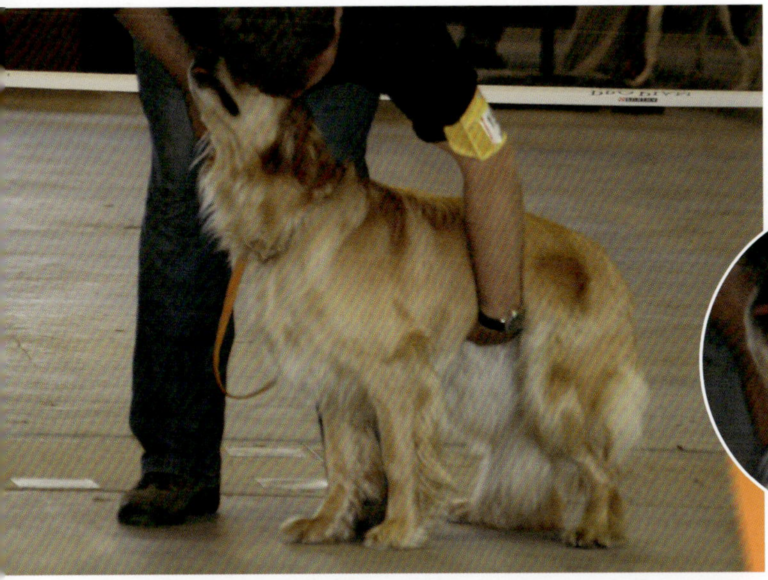

His handler is trying to pose the dog for the judge and uses his hands to help him into position. This manhandling, which is accompanied by the handler leaning over the dog, is too much. The dog licks and tucks his tail between his hind legs.

His mouth is now closed, with the corners of the lips pulled back. His eyes are very small and he slightly pulls his ears towards the back. He continues to carry his tail between his legs showing that he is far from comfortable.

Taschka (*left*), and Arwen (*right*) tuck their tails while vomiting.

This former street dog is feeling insecure, which is evidenced by her tucked tail. Matters are made worse by the tight leash, which means she cannot avoid the situation. Former street dogs often struggle in this way, even when they are being treated well.

Tucked tail while insecure

- Tail pulled under the belly

Additional clues indicating tension:

- Rounded back
- Hunched posture (crouching)
- Ears flat and pulled back
- Licking

8. THE COAT

The texture of the hair, whether it is dense or soft, short or wiry, long or curly, depends on the dog's genetic make-up which he received from his parents. To gather information about a dog's coat we look at direction of growth, texture (many illnesses are reflected in a change of the hair), change in colour and changes that could be the result of an emotionally stressful situation.

In 2007, a team of scientists attributed two features to an associated gene – the colour white, and the ridge along the back which is typical in Rhodesian Ridgebacks. Both traits carry a great medical risk factor. Many white dogs experience hearing loss, and dogs with a ridge suffer from a neural ear defect known as dermoid snus. There are other colours, such as the merle, that under certain genetic conditions, also carry a predisposition for illnesses. First time dog owners should also be aware that fur and coat colour are related to the physical health of dogs. This closes the circle to the emotions, which are also influenced by pain and physical ailments.

When a dog has an excessive amount of hair on his face, it is almost impossible to get any information from him. He is not only difficult for us humans to read, but for other dogs as well.

A new haircut allows this dog to see and be seen and understood more clearly by other dogs.

If you don't like the look of a straight cut above the eyes, you can use thinning scissors to thin out the hair over top of the eyes and along the cheeks as we did with Tamika (*left*) and Curly.

These types of hair clips are very popular among the owners of small, long-haired breeds. The idea is to allow the dogs to see better, which is commendable. However, if the hair is pulled too tight the dog may experience tension on the top of his head, and may develop headaches.

Prior to being clipped, Ischka's eyesight is compromised and it is hard to read her expression.

Now clipped, the difference is tangible.

COAT CARE

Depending on the texture of the hair, long hair generally requires more care than short hair. It is a good idea to get a young puppy used to gentle, respectful brushing, even if he doesn't really need it. Turn your grooming session into a wellness treatment by grooming slowly and gently with soft brushes. As a dog ages and loses mobility, he may not be able to groom himself as effectively and may need more assistance. Long-haired dogs require more attention. Matted hair between the legs, in the groin area, and around the elbow and chest can cause pain and limit movement. This can have an effect on the dog's behaviour. Twigs, burrs and thorns can get tangled in the coat, changing posture and compromising the dog's ability to move, which will influence his emotional outlook.

PILOERECTION

When we think about raised hair, we usually have a picture in our mind similar to that of Rani (*below*) with all the hair on her back standing straight up. Many people connect this look with insecurity or even aggression.

The raised back hair is also called piloerection and can be compared to us having goosebumps. How goosebumps are created is not really known, but there are a variety of causes, such as cold temperature, and positive or negative excitement. Just think about a piece of music that has really touched you, a caress, or the sensation of the 'hair on the back of your neck standing up' because something was spooky. You may experience the same sensation if you sense someone is secretly watching you, or if you feel really angry.

Very excited – surprised

- Hair raised from the back of the neck to the tail

Additional clues indicating tension:

- Tail held higher than the back and moving
- Topline falling forward
- Joints bent
- Body hunched (crouching)
- Ears pulled back
- Mouth open with a crooked grin
- Looking at the object
- Hesitation while moving forward

BEFORE: Earlier, Rani, the Golden Retriever, saw a stack of wood and other dumped items in the forest that had not been there before. She was startled at the new find and became very excited.

Maybe you and your dog have been in this type of situation? You are in a familiar area and, all of a sudden, you see something new. For some dogs, a startle response can be triggered by bag of rubbish, for others it could be a tree stump looks different after a fall of snow. A longhaired dog that had been clipped can, momentarily, cause confusion among his dog friends. Primarily, the dog is showing some degree of excitement. He may be insecure, fearful, angry or nervous but, equally, he may be perfectly happy and simply showing his excitement.

AFTER: Following her startled response, Rani has settled down noticeably. The raised hairs on her back are almost flattened and she is able to turn away. However, she is not feeling completely safe in the situation.

- Hair on the back is not so raised

Additional clues indicating tension:

- Posture still hunched
- Tail carried low
- Long mouth
- Ears held towards the back

EXCITEMENT

About playing

Some dogs are always excited when they see other dogs or get to play. They show their excitement by raising the hair along their backs.

Quivive, the Pyrenean Sheepdog, and her Beagle-cross friend, Coco, have known each other since they were puppies and love to play roughly with one another. When Coco is happy and excited her back hair always stands up (*pictured right*).

EXCITEMENT

Playing with increasing insecurity

Rani, the Golden Retriever, plays with Spot, the Border Collie mix, a dog she knows well and likes. The two of them start to play roughly and, as the game progresses, Rani becomes more insecure. The raised hair along her back shows her excitement, maybe her joy, as she plays with her long-time friend.

But as the game gets wilder, Rani shows her insecurity:

- Tail lowered
- Hair raised around the shoulder area and base of the tail
- Topline sloping
- Strong bend in her knees and elbows
- Hair over her shoulders slightly raised
- Back rounded towards the back
- Tail carried low

To slow Spot down, Rani makes herself smaller. Her body is turned away from Spot. You can see raised hair all along her back.

Rani sniffs the ground as a means of calming Spot. Her topline is very rounded towards the back and she is carrying her tail low. It looks as though Spot was planning to jump towards Rani, but understood her plea for a break at the last second and aborted the playful attack in mid-air. Rani's back hair is flat again.

EXCITEMENT

With a little more insecurity

Rani meets Kissy's puppies for the first time. She is a little freaked out, but also curious. After checking out the pups she settles down. The hair on her back starts to flatten and her ears are further forward, towards the puppies. She also shifts her body forward. Her meeting with these bold puppies is polite and amicable.

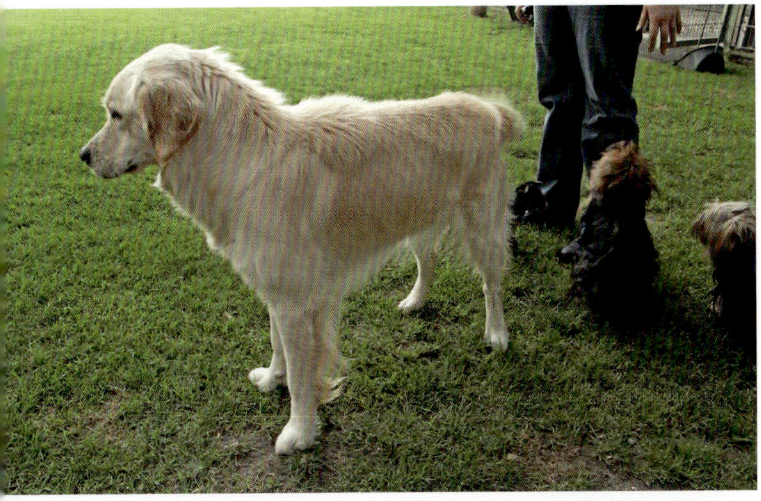

Rani's initial response.

- Hair raised just behind her head all the way to the tail

Additional clues indicating tension:

- Tail level with the backline and moving
- All four legs braced
- Neck raised
- Tension in her face
- Point of balance over her feet

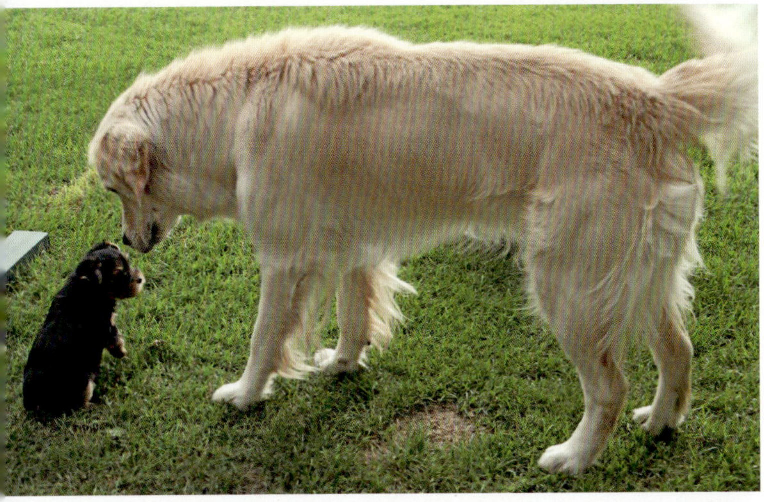

Rani dares to approach the puppies. She shifts her weight towards the back and her hair is still raised.

After the first contact Rani's hair becomes noticeably smoother. She has now recognised that the small creature is actually a dog ...

PROTECTING PUPPIES

Puppies and young dogs can be irritating to adult dogs, especially when they haven't had any experience with young dogs in the past. Not every dog approaches puppies as carefully as Rani. Depending on the older dog's disposition, age, pain level or physical ailments, this type of interaction can easily get out of hand and be dangerous for the puppies. The puppies are usually safe and protected when they are within their own family, but this instinct to protect is not guaranteed. Indeed, some mothers will hurt their own puppies, so you cannot rely on this instinct. The same applies to people; you assume that children are safe within their own family, but sadly that is not always the case.

EXCITED

Annoyed

Rani, who is pictured between Tamika in the foreground and York, a light-coloured Golden Retriever, in the background, is a very patient dog. It takes a lot for her to get really angry and be on the offensive. The behaviour shown in these photos is very rare for her. On this occasion, Tamika, must have really annoyed her. You can see this in the way she holds her ears, her eye contact and the wrinkles in her face. She shows her teeth for a split second. York (*in the background*), a light-coloured, male Golden Retriever observes the altercation. He is not involved but he is emotionally affected by the situation, evidenced by his pulled back, flattened ears, his tightened mouth and the beginning of a crooked smile.

Rani

- Hair raised from the neck to the middle of her back
- Ears pointing forward
- Fixed eye contact towards Tamika (in the foreground)
- Corners of the mouth pulled up
- Upper canine incisors exposed
- Clear wrinkling on top of the nose
- Point of balance forward

Tamika understands and turns away. She has rarely seen Rani this upset. Rani's face relaxes right away and she stops her forward movement. The raised hair on her back shows she is still exited. Also note York watching what is going on in the background. When the tension decreases he blinks, either to calm himself or to send a clear signal towards the two females.

HIGHLY EXCITED

There are other reasons why we see a dog's hair raised.

Quivive has seen or heard something that has given her the goosebumps. The long hair on her neck and shoulder region are clearly standing up.

This Tunisian street dog probably has puppies and is therefore very concerned when the white dog, a stranger to her, appears. Her tail is raised high, her ears are pointing forward and her mouth is short (*see inset*). All these signals show her determination to protect her puppies at all costs.

FACIAL HAIRS AND THEIR MEANING

The whiskers (also known as vibrissae or sinus hairs) are located on the dog's face: above the eyes near the eyebrows, below the bridge of the nose on the upper lips, under the chin and behind the corners of the mouth. They are stiffer and harder than ordinary hairs and can each be of different length or strength. They can be white or black and are easier to see in short-haired dogs.

There are also a few whiskers under the chin.

This dog's whiskers are barely visible, but the grooves where the hair shafts grow are very clear.

Whiskers stand up when a dog is excited, and the excitement can be caused by fear, anger or an attack, but we also can't rule out pain. Thus the whiskers provide another/additional source of information for other dogs on how an individual is feeling.

Here the Golden Retriever Rani (*right*) is getting attacked by a Husky. She is very afraid. The whiskers on the side of her nose are raised on her face, and also on the face of the attacker.

The whiskers that grow in the area of the eyebrows rise up when the dog pulls his eyebrows together and therefore help to express his mood more clearly.

Neutral position of the eyebrows (soft eye). The whiskers around the eyes are slightly raised.

The eyebrows pulled towards each other (fixating). The whiskers are pointing forward.

Beside the whiskers, there are some additional hairs on the upper and lower lip that are different in texture from the dog's regular fur. They may also be raised if a dog is excited or in pain. These hairs are a little bit thicker and longer than the regular hairs around the mouth, but much shorter and thinner than the whiskers.

IMPACT OF STRESS

Did you know that dogs can turn grey from stress?

This male Rottweiler used to have the black and tan markings that are typical of the breed. Unfortunately, his guardian had to re-home him for two months and, instead of living in the house, he was chained up outside. When he was able to go back home, he had turned almost completely white on his face and on the lower part of his body.

9. PAIN AND AILMENTS

Why is there a chapter about pain and ailments in a book about emotions? Well over 80 per cent of the dogs who have shown a change in emotions experience pain, or are suffering from some type of chronic illness. Dogs are masters at hiding their pain and can compensate for a long period of time. As their guardians, usually the first thing we notice is a change in behaviour. A dog may be uncharacteristically moody or aggressive, or he may appear to be lazy and require more rest. He may also start to shy away from touch, become noise sensitive, or avoid other dogs. Dogs will often communicate pain and discomfort through yawning or by scratching themselves.

Pain is often experienced in the back, and this may be seen by a changed gait and posture. Other ailments may present through subtle changes in the way a dog carries his ears or tail, or he may have a different facial expression.

Of course, we can only cover a very small section of possible ailments in this book. But we would like to use this chapter to help you train your eye so you become aware of the subtle changes. If we recognise that a dog's behaviour can change due to an illness, in addition to sharing this valuable information with the vet, we can better understand their emotional state, develop empathy, and protect the dog from unwanted contact. This makes it easier for humans and dogs to live together during an illness and deepens the bond through mindful handling while the dog is unwell.

PROTECTIVE POSTURE

A dogs' protective posture is not always easy to recognise. When it is as obvious, as in this photo, the dog is in severe pain.

Tiber developed a spondylosis in his advanced age. While in an acute state he no longer allowed his back to be touched and barely dared to move.

- Head carried low
- Topline straight
- Increased yawning

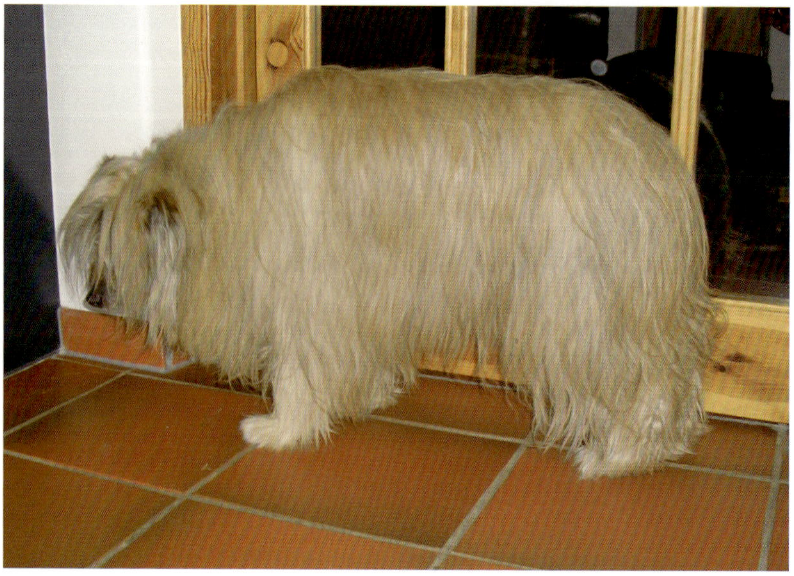

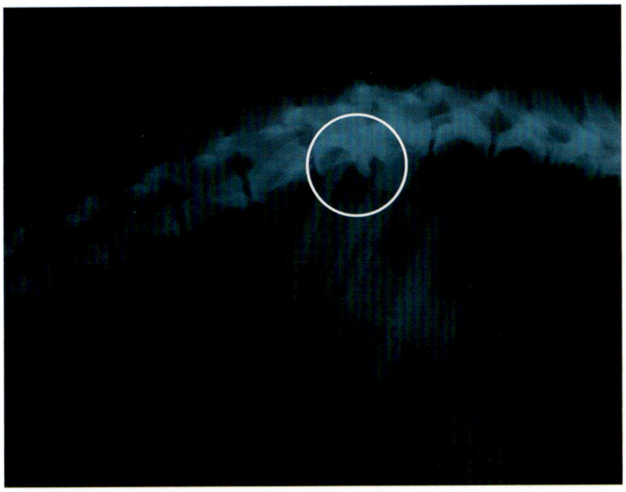

The X-ray clearly shows the developing bone spur on the affected vertebrae (*see circle*).

In addition to pain medication prescribed by his veterinarian, wearing the Tellington Body Wrap helped during the early phases of the pain. We added a TTouch Body Work session, as well as regular chiropractic and osteopathic treatments, to his regime to help the rest of the vertebrae and back muscles to stay flexible.

SEVERE PAIN

Due to back pain

Aside from a change in posture, severe panting is a sign of pain.

Quivive, the Pyrenean Sheepdog, suffered from severe pain in her back after a fall. Here she is panting due to her pain. After a visit to her veterinarian she was treated with pain medication and chiropractic and osteopathic adjustments.

- Tail clamped and held down stiffly
- Topline extremely rounded
- Ears pulled back
- Corners of the mouth pulled back

SEVERE TENSION

Due to stiff back

This female Weimaraner is holding her back almost straight and the same goes for her belly. It seems she is trying to avoid movement in those areas. Her gait is stiff and hesitant and her tail seems not to be moving. In this case there could be a stomach or gut issue, evidenced by the excessive drooling.

We could observe threads of saliva the entire time we were watching this Weimaraner. When meeting other dogs she was indignant and angry. We don't know what was wrong with this dog, but she was definitely unwell.

SEVERE TENSION

Due to pain in the belly area

Sometimes a dog may display a protective posture, indicating pain, which will lead us towards the wrong diagnosis.

This is Quivive while she was suffering from acute pancreatitis. She was treated for back pain and her suffering was increased due to the wrong diagnosis. Her blood work was in the normal range and she did not display the typical symptoms of pancreatitis. When the pain became unbearable, her entire body would shake and she could no longer lie flat on her side. She preferred the position you can see in this photo, where she could raise and support her neck. Her back was usually curved and her hind legs pushed forward and held very stiffly.

ASYMMETRIC EAR CARRIAGE

Painful conditions can also mean that one ear is suddenly held differently to the other.

Quivive's normal ear position is a half-raised tipped ear, which is clearly seen in this photo.

Due to a painful abscess in her left eye socket, Quivive is carrying her left ear towards the back and it is more raised at the base than the right.

Quivive was not in pain when this photo was taken, and you can see the hairs around her nose are flat

While she is in pain, the hairs around her nose are raised, and the area appears to be enlarged or swollen.

RAISED FACIAL HAIR

Pain can also cause the hairs around the nose, the whiskers or vibrissae to stand up.

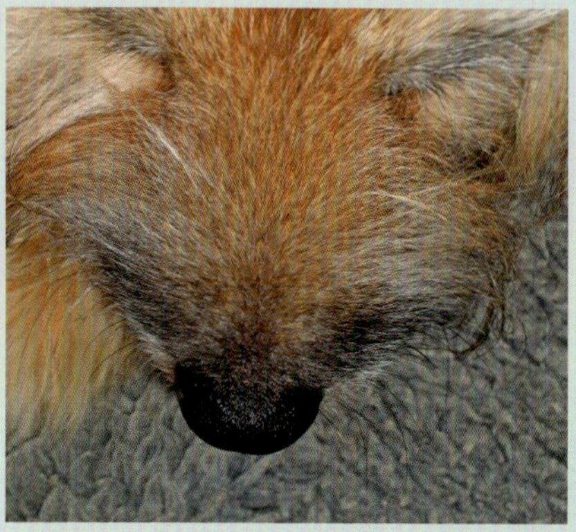

BEFORE: Tiber was in pain when this photo was taken. The short, black hairs around his nose are raised and the hairs underneath are not easy to see. His vibrissae are also raised.

AFTER: This photo was taken 18 months' later and the formerly black hairs beside the bridge of his nose are now grey. Even though the perspective and angle are slightly different, you can see that the short grey hairs are lying flat and the white hairs below the nose are clearly visible.

CHANGED GAIT

FOLLOWING TRAUMA

A change in length of a dog's leg can lead to muscle atrophy in the leg that is used less, and to overloading of the leg that is used more often. Since dogs are four-legged, they also tend to over-compensate via the uninjured parts of the body (in this example via the back and forehand).

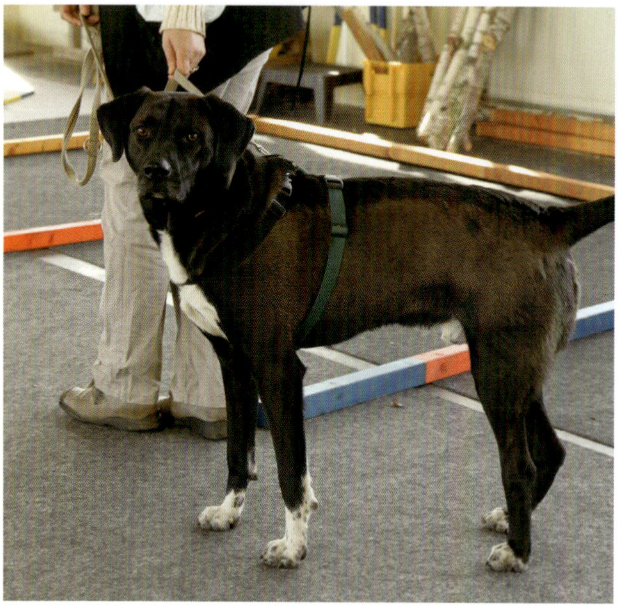

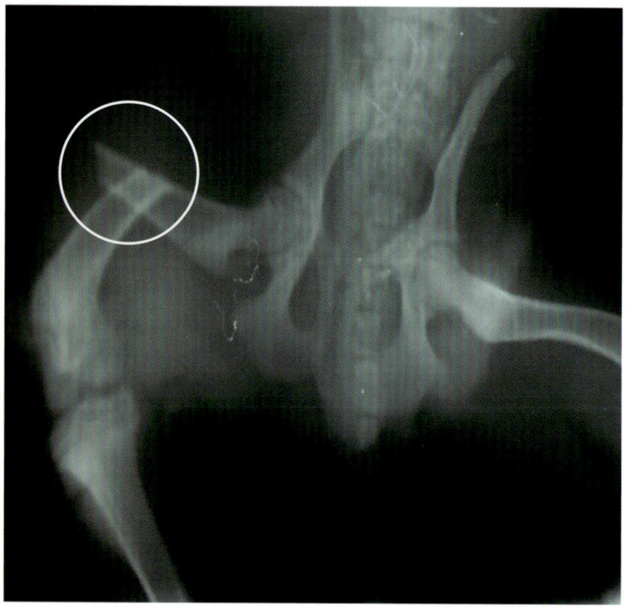

When he was young Bonnie was involved in a car accident and suffered a severe femoral fracture of his left leg (*see X-ray, pictured right*) that had to be set with a screw. He received excellent care, including physiotherapy and while he improved greatly, he did not weight-bear on the shortened leg.

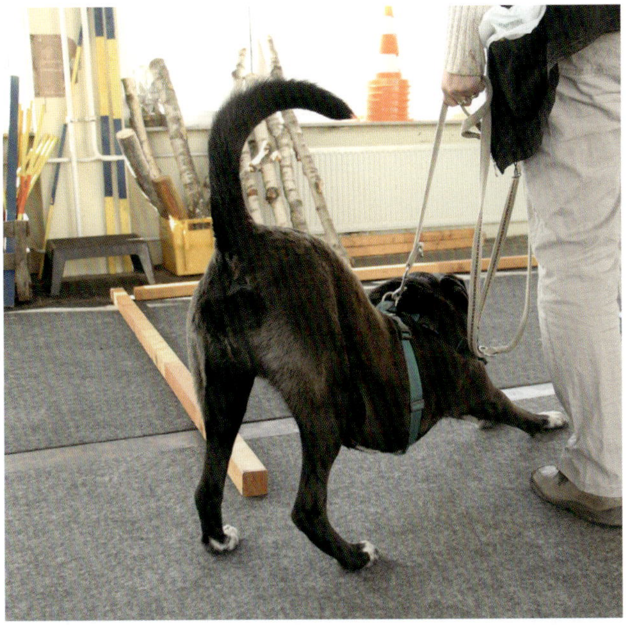

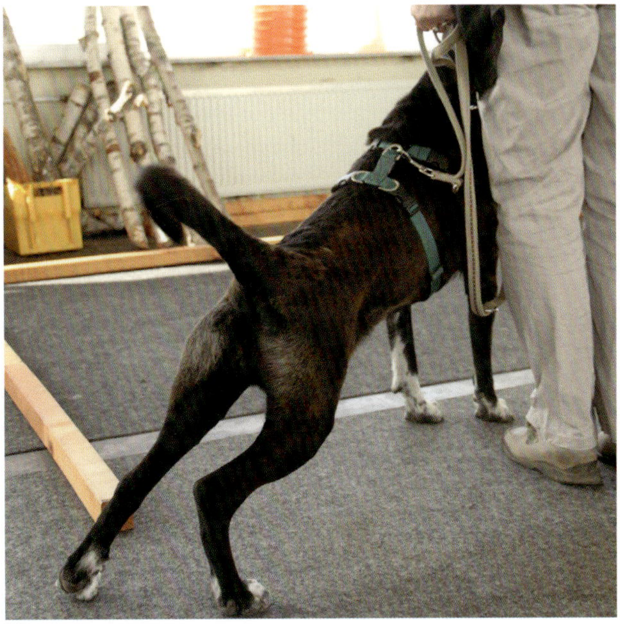

These are relief postures for the back, which compensate for the uneven load on the hindquarters. Bonnie uses these postures very often.

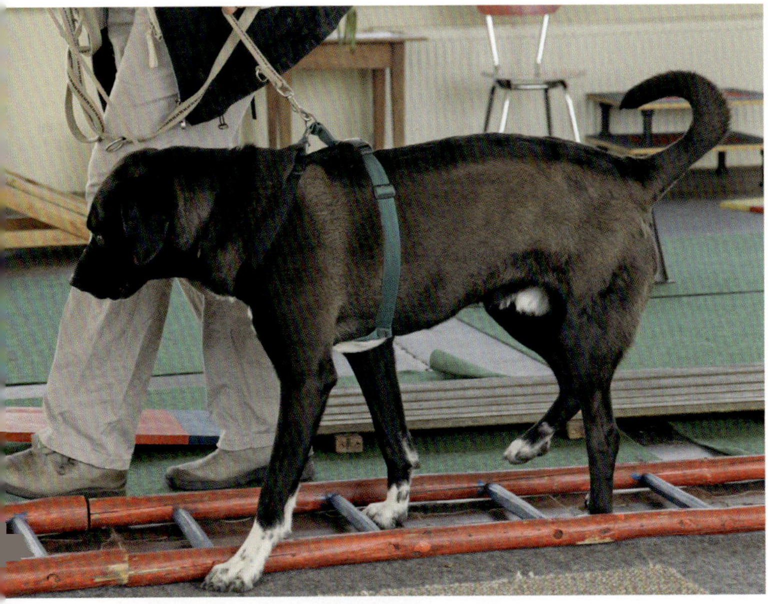

At the start of training, it was difficult for Bonnie to walk through the ladder in a straight line.

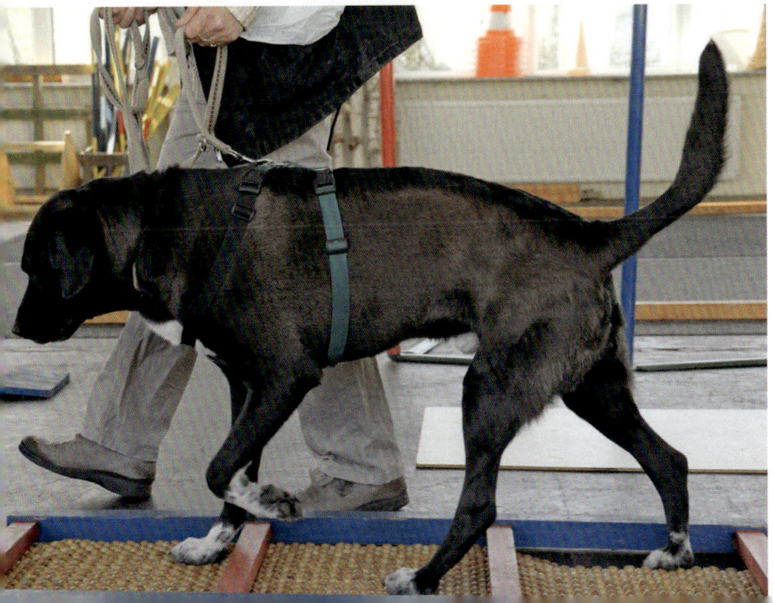

Using an obstacle course during the Tellington TTouch® sessions, he was quickly able to master most of the obstacles while actually using his shortened leg.

Bonnie was able to keep his back straighter and used relief positions less frequently. We kept the sessions very short in the beginning, never working more than five minutes, because he had a very short attention span and his body tired easily. With additional training we gradually increased the challenge.

ARTHRITIS

Dogs suffer just as much from arthritis as people do. Subtle changes in their gait go hand in hand with pain, but are often difficult to recognise. A change in behaviour may be the first indication.

Arthritis can be caused by faulty conformation, it can be hereditary, or it may be the start of the ageing process.

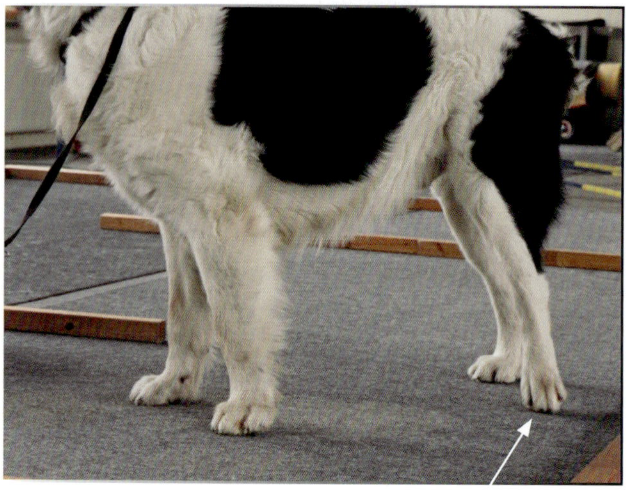

The 11-year-old Landseer male, Anatol no longer flexes the joints of his left hind leg due to the stiffness brought on by arthritis in his tarsus. His body seems to have 'forgotten' the injured leg, and he drags the left back toe on the ground when he walks (*see arrows*). This, of course, causes problems in the long run from friction to the top of the paw and the nails. The muscles of his hind end have already atrophied noticeably.

Initially, Anatol could only step over flat obstacles. If he attempted a raised pole his left hind leg became stuck. Even at this advanced age (for a Landseer), we can still do a lot to improve functionality and mobility, and thus enhance the general quality of life.

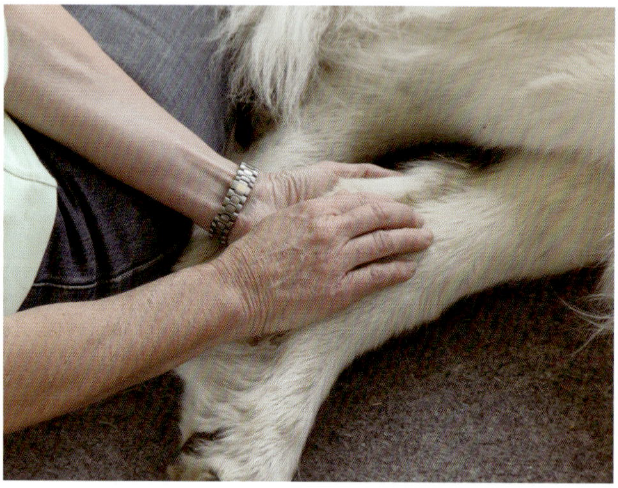

For Anatol, the Tellington TTouch® and ground exercises brought a lot of relief. He was able to use his hind end much more effectively and shifted his weight off his front end. Anatol's guardian worked with him diligently and gave him all the tools to provide the best quality of life possible. He lived until his twelfth year.

INSTABILITY

Sammy, an eight-month-old Elo (a breed recently developed in Germany from Old English Sheepdog, Eurasian, Chow Chow and Spitz type dogs) came in for training as he would not stop barking at the dog park.

He became a star student, always eager to participate in any exercise and he loved puppy obstacle training. Then, after four months, his behaviour changed. He refused the obstacles, didn't want to play with the other puppies, and barked non-stop during obedience exercises. Neither his trainer, nor his guardian, could explain the change in behaviour.

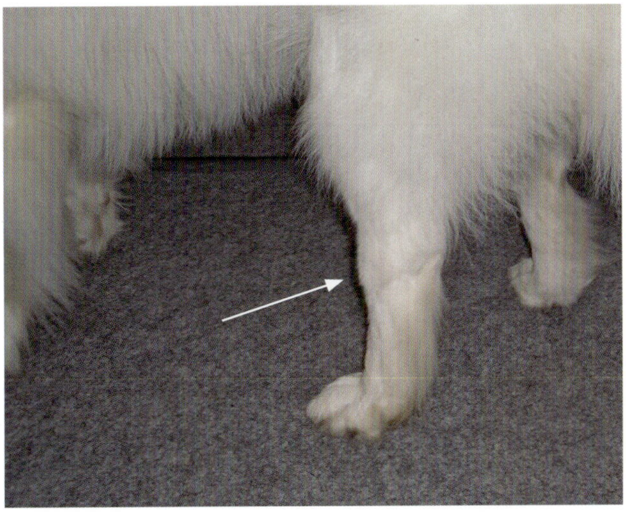

Under careful observation, an instability in both tarsal joints became noticeable to a point where it was over-stretched even when Sammy was standing still. After investigation it became clear that Sammy's change in behaviour was due to the instability of his tarsal joints.

This is the normal angulation of the tarsal joint.

ASYMMETRICAL CHANGES

Asymmetrical changes of the hair at the bottom of the paw, the pads of the paw and the nails indicate uneven wear. This could be caused by age-related changes or a very subtle, barely noticeable change in gait or carriage to protect the body.

CHANGES IN COAT AND SKIN

Teary eyes and obsessive licking can change the colour of the coat. Pressure points from a poorly fitted harness can have the same effect.

Note the change of colour around the mouth, probably caused by excessive licking.

Colour change below the eyes resulting from continuous tearing of both eyes.

Changes in coat colour such as this, can be the sign of something that is going on inside the dog's body, so it should be investigated.

In this case a urinary tract infection was implicated, and once it was cured, the hair discolouration changed back to its original colour. It is interesting to note that the colour change is right on top of the bladder meridian.

Partial change in coat colour can have many different causes, ranging from the undercoat failing to shed to a series of illnesses, or a nutritional deficiency. There are some cases where the cause cannot be found.

Baja, an Appenzeller, has developed a brown hue on her coat. The cause is unknown.

This street dog in Nepal is showing partial brown colour changes on the entire body with very rough, dull hair.

The coat of this Tunisian street dog is rough, thin and dull. This is a clear sign of malnutrition and lack of nutrients, most likely combined with worms or parasites. This male is a little excited, probably because he has seen Katja with her camera. The hairs on his back are raised. He wants to take the food he has found to a safe spot.

Organic ailments, such as thyroid imbalances can also cause a dull coat with hairs that stand up. When the thyroid underperforms you will often see a 'rat tail' with sparse hair growth, similar to the dog in this picture.

Parasites such as worms or mites (sarcoptes), or illnesses of the internal organs that remain untreated, can cause extreme changes to the skin (elephant skin) as well as the coat, as you can see in this street dog in Nepal.

Many dogs suffer from allergies that cause changes in the skin. Parasites can also create havoc with the skin and coat. Dogs that are constantly itchy, or have inflamed skin, are often nervous and easily agitated.

A hairless Chinese Crested with obvious skin problems.

This English Bulldog has red skin with little hair.

SWIRL DEVELOPMENT

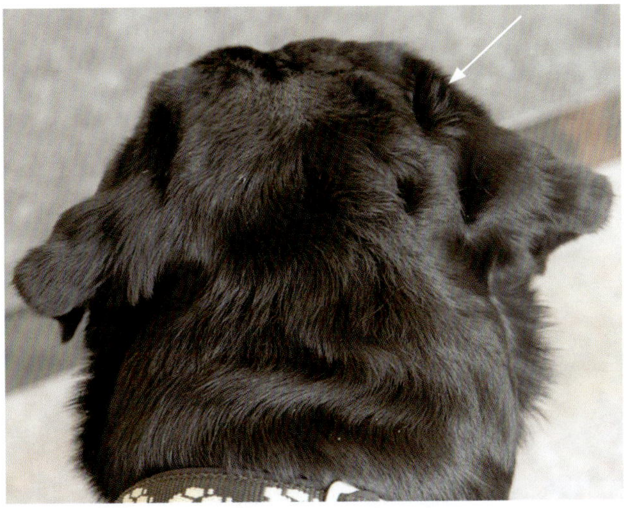

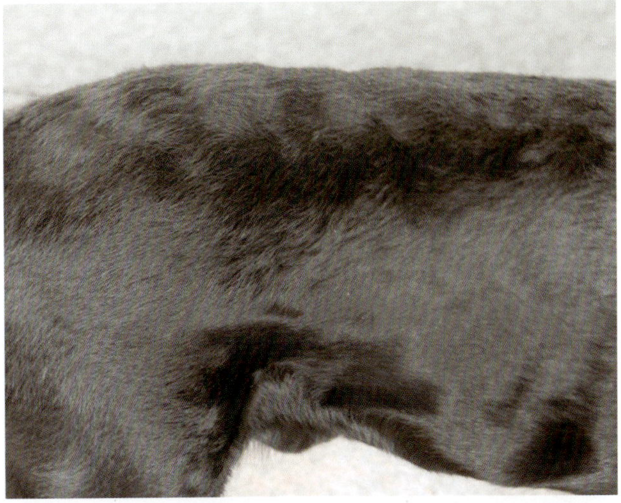

Mo, a male Labrador, displays a noticeable and asymmetrical development of swirls beside the base of his right ear (*see arrow*). Towards the back, the hair coat is rougher and stands up more compared to the rest of the coat over his shoulders and the front section of his back.

Mo was very reluctant to do the obstacles in the Tellington ground exercises. He avoided all obstacles that moved and was afraid of many of the other obstacles, even though he had been familiar and comfortable with them for years during his performance in dog sports. However, in agility the obstacles were always taken at high speed, which allowed him to hide his concern. Mo was overly sensitive to noises and touch.

A chiropractic examination showed that, probably due to a fall off the dog walk in agility, he suffered from massive and painful tension along his entire spine. His pelvis, shoulder and vertebrae in the neck were also affected, as well as the facia along the entire back. Through chiropractic treatments and Tellington TTouch® sessions, his caution around the obstacles slowly disappeared and his noise and touch sensitivities were greatly reduced.

As Mo was feeling better, the swirls along his back disappeared and the hair in that area returned to being soft and flat. Mo's guardian stopped his agility training and concentrated on leading him slowly, in balance, through the Tellington TTouch® obstacles.

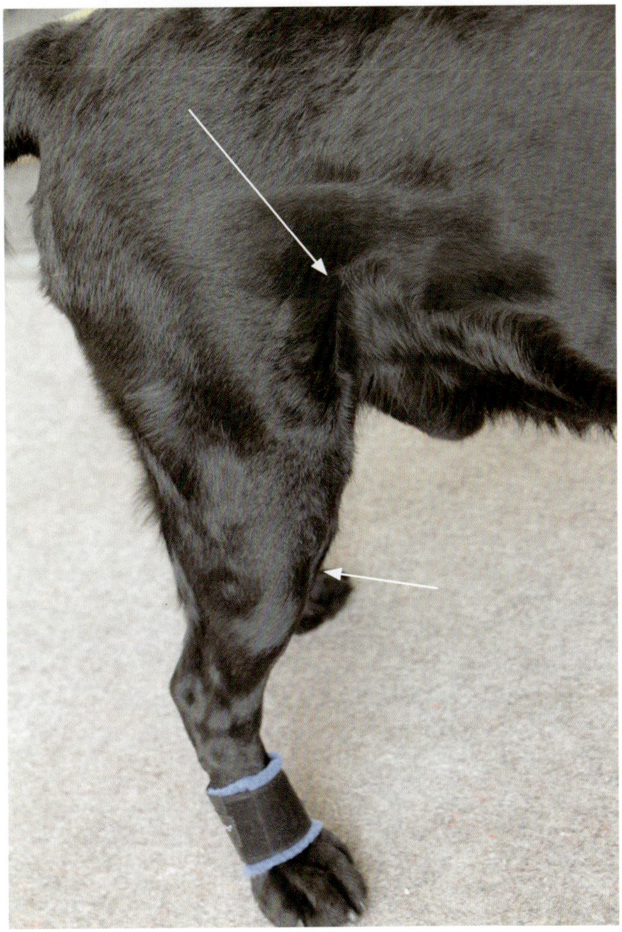

The hair started to curl from both knee joints up into the flank (*see arrow*). Here Mo is wearing a velcro ankle bracelet, a Tellington TTouch tool to bring awareness to his hind feet and improve gait.

SUB-CLINICAL HYPOTHYROIDISM

A change in the texture of the coat can be one of many non-specific symptoms, such as lack of impulse control with aggression and panic, touch and noise sensitivity, inexplicable fear, interruption of the female's cycle or male's potency, or food issues, to name a few. These can all be clues leading towards a diagnosis of hypothyroidism.

As opposed to the clinical form (starting with the coat losing its shine, graduating to excessive hair which later turns into bald spots and a rat tail, diet resistant weight gain, and depressed behaviour), the coat often seems sparse with a lack of undercoat. Sometimes the hair is dull and coarse, but the exact opposite can also be the case where the hair is smooth and shiny. The dog can be extremely thin, even though he is taking in a lot of nutrition.

Momo, a blue and white female Border Collie, was deaf and suffered from a severe case of hypothyroidism. She was very hyperactive and lived in five different homes in her first year until she found a forever home with Katja.

Her previous guardian kept her in a small cage which resulted in many stereotypical neurosis which afflicted her for the rest of her life. In spite of much effort (western medicine and alternative treatments) Momo was a dog who rarely looked happy. Her coat was always dull and lacked colour.

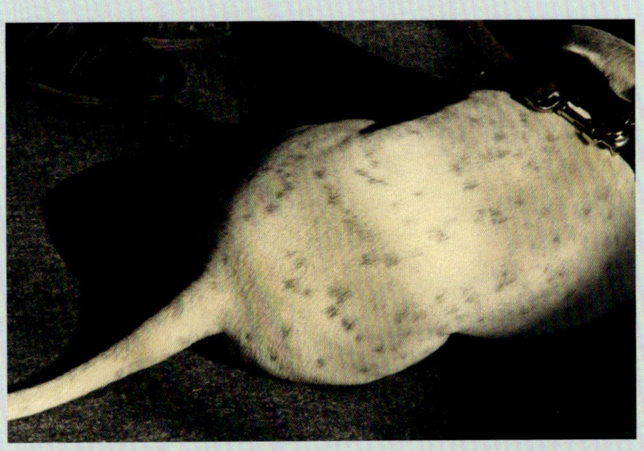

BONNIE: A CASE STUDY

Bonnie presented with extremely impulsive behaviour which could quickly switch from excited to aggressive. She would then use her teeth. Her coat was sparse; you could look right through it and see her skin.

In the beginning Bonnie could only be touched by her own people. Throughout the Tellington TTouch® sessions she became less impulsive and, with some food distraction, Gabi was allowed to TTouch her and she relaxed. However, she hit a plateau and her behaviour did not improve any further. Gabi suspected that sub-clinical hypothyroidism could play a role in Bonnie's behaviour. However, this could not be confirmed by a blood test because the only way to draw blood from her would have been under anaesthesia, and her guardian did not want to put her through this ordeal.

The sub-clinical hypothyroidism seems to be quite similar to the human disease, Hashimoto-syndrome, an autoimmune condition where the body's cells destroy the thyroid. This disease is hereditary and can stay undetected. Certain medications and an on-going, increased flooding of the stress hormone, cortisol, can trigger an outbreak of sub-clinical hypothyroidism. (Bettina Zimmermann: *Thyroid and behaviour*).

Many veterinarians focus on the presentation of clinical hypothyroidism and ignore sub-clinical hypothyroidism because the applicable blood values are still within the normal range. This increases the difficulty of a correct diagnosis. In addition, there are countless ailments that present with similar physical symptoms and behaviours. When dealing with dogs who present with behavioural symptoms, it is advisable to work in co-operation with a veterinarian, and a veterinarian who specialises in behavioural medicine.

For the guardians of dogs suffering from sub-clinical hypothyroidism, the road to finding a diagnosis is often a very long and exhausting. They are often seen as ineffective trainers when the behaviour problem is actually based on the dog's illness. In most cases, help is initially sought for the behavioural problem before a health condition is implicated.

Hypothyroidism can be found in all breeds, as well as mixed breeds. To date, there has been very little research into the conditions and the problems it can cause.

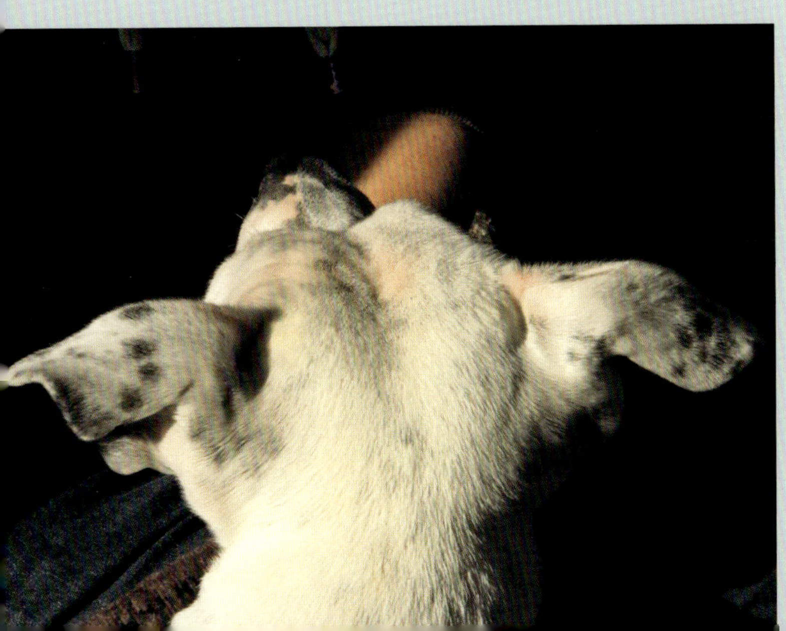

This photo shows that the hair growth is quite sparse on top of the head where Bonnie is starting to wrinkle the skin. You can see her pink skin shining through the coat. The hair is thick and normal from the base of the ears going towards the back.

ANAL GLANDS

Many dogs suffer from blocked or infected anal glands, which should empty themselves when the dog moves his bowels. If the fluid in the anal sacks thickens, the sac can become blocked and cause a build-up of fluids. Many dogs respond well to homeopathic treatment for this ailment.

Besides a predisposition for this problem, there can be a number of different reasons why a dog may suffer from an anal gland issue. It is easy to diagnose – the dog starts scooting his bottom along the ground to try and relieve the pressure of the glands. Another symptom is continuous licking of the rectum to ease the itchiness.

In a more severe case, the dog may shorten his stride when walking. This usually starts very subtly and when the dog gets to a point where he is noticeably lame, he will have a massive problem with his anal glands. Other dogs develop a touch sensitivity on the hind section of their back, which can have many different causes.

In some dogs you may notice a redness below the tail and the is often carried in an unusual manner. If the infection continues, the dog will experience a rise in temperature and general lack of well-being. An untreated infection of the anal glands can turn into a very painful abscess that will need he attention of a vet. Itchiness of the rectum can also be a sign of parasites.

EYE CONDITIONS

Dogs can suffer from a wide range of genetic and acquired eye problems and a change of behaviour is often the first, or only, clue that something is wrong. If you suspect an issue with the eye, you need to consult with a vet right away and, ideally, one who specialises in ophthalmology. Just like people, dogs may need immediate emergency assistance when the eye is affected – in the case of glaucoma, for example – to protect their vision.

CLOUDY LENSES

Age-related clouding of the lens in Taschka

There are some ailments of the eye such as cataracts and progressive retinal atrophy (PRA), that will change the colour of the pupil or cause the pupil to be permanently dilated and therefore give an impression that the eye is fixated. Since the dog's vision is impaired, he may accidentally create direct eye contact since he now relies increasingly on his sense of smell and sees the world with his nose. However, the unintentional eye contact can easily lead to misunderstandings with other dogs.

PROGRESSIVE RETINAL ATROPHY (PRA)

This ailment is hereditary; there is no cure and it will lead to blindness. In daylight you will see a strong reflection of the back of the eye, as can be seen in the following photos:

Tiber pictured at three years old, and already completely blind from PRA. His eyes literally glow. The iris is no longer visible due to the full expansion of the pupils.

Tiber now aged 10. Note that the lens of the left eye is starting to get cloudy.

In comparison: Quivive's left eye (*pictured left*) is clear. The right eye (*pictured right*) is tearing due to a mild case of conjunctivitis. The red conjunctiva is not always visible from the outside, therefore it is important to consult a vet if there is consistent tearing of the eye. Invasive eyelashes, foreign objects, or a scratch to the cornea could be the cause, and could lead to moodiness as well as physical discomfort.

NEUROLOGICAL AILMENTS

A dog can suffer from a large variety of neurological issues. In ageing dogs, vestibular syndrome is not uncommon. It is often misread as a stroke, which is much more rare in dogs.

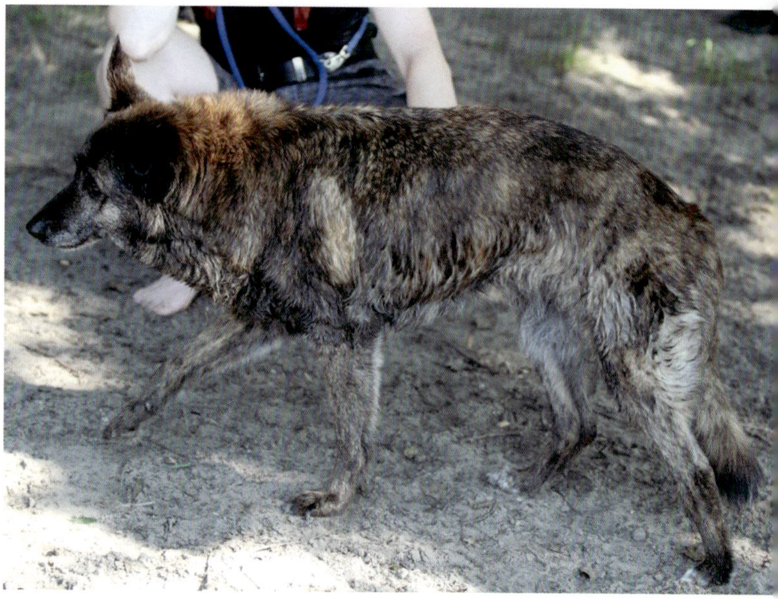

The head and ear carriage, as well as an ataxic (un-co-ordinated) gait, indicate a high probability that this dog is suffering from a neurological ailment or vestibular syndrome.

Quivive showed a lot of symptoms of vestibular syndrome as she reached an advanced age. She would be disoriented and experience disruption in her co-ordination, which included tipping her head and ears, and the midline of the tongue was no longer straight.

A crooked face, which is probably due to a stroke. The dog was restless and showed a change in his awareness and in the way he interacted with other dogs.

There are a lot of other reasons why a dog is no longer able to hold his head straight. If this happens suddenly, and is on-going, consulting your vet is highly advisable.

TRISTAN: A CASE STUDY

Bacterial meningitis

When Tristan was two-and-a-half years old, he contracted bacterial meningitis from an unknown source. His guardians and vet fought for weeks to save his life. Since the disease was diagnosed quite late, the back half of his body was paralysed. From that point onwards, Tristan has received daily Tellington TTouches®, plus a lot of other support. He attends swimming and physiotherapy, has regular treatments from an osteopath and acupuncturist. and takes homeopathic remedies.

Tristan never lost his will to live. The disease marked him, but as soon as he got through the worst of it, he was happy again. He did not seem to care that he could no longer walk. He 'smiled' when he greeted people and played with his canine and human friends lying down. Luckily he was able to swim fairly soon. He can now walk, manage some stairs, and even wag his tail. His guardians never gave up on him. Our dogs are good at teaching us that it is not the situation we find ourselves in that makes us unhappy, it's what we do with it.

Tristan's face before the illness.

Tristan while he was sick. The paralysis affected his body and, within a few days, the musculature on his face was also diminished.

Tristan in recovery: The muscles in his face have rebuilt themselves.

Before: Tristan before he got sick.

After: Tristan now enjoys a good quality of life and is able to join his family on walks. When he moves he has a noticeable wobble because he cannot properly co-ordinate his left hind leg.

AGEING DOGS

Ageing with dignity applies just as much to animals as to ourselves. As dogs age, they require special attention and care. Many tasks that are natural for the young dog now require more energy and are more challenging. Since the ageing process is often slow, it is easy for us to miss that the dog's senses are diminishing, that his heart may not be so strong, or that his joints are aching.

There is a lot you can do to maintain the older dog's physical wellbeing and his joy for life – after all, your dog is only as old as he feels!

Constant pain should not be the fate of an old dog. Besides alternative healing methods, he can be treated with western medicine and plant-based pain medication. Even old dogs experience joy when they have a good roll. If the dog is still able to do this, it is the best exercise for keeping the back flexible and healthy.

Swimming is gentle on the joints at any phase of life. However, racing after a toy and bringing it back – no matter if it is in the water or on land – is very hard on the dog's body due to the added stress on the joints and tendons.

When she was young, Eve loved to retrieve.

She still enjoyed the game when she was 15, even though her senses and awareness were no longer optimal. Her search and retrieve radius shrunk from 45m to about 1m.

Taschka roughly one year old.

Here she is at 15 – you can clearly see her change in posture

LIMITED MOVEMENT

Some older dogs show a change of gait due to physical limitations. The topline is often very straight and the musculature of the back can become very tense and rigid. The joints get stiffer, just as they do in humans, and moving becomes more difficult. The tail is often carried low and is limited in its movement which, of course, affects the dog's balance. Sniffing the ground becomes more difficult because the shoulders and elbows don't bend as well.

Older dogs will often find it easier to pace (lateral gait) instead of trot because there is limited movement in the back.

THE PROCESS OF AGEING

As dogs age they experience changes in their bodies. They become less mobile and cannot react as quickly. Some body parts are difficult to reach and clean, and pain may be a constant companion. No dog should experience chronic pain; your veterinarian can help with this.

Age related limitations can be frightening and cause a dog to become very insecure. An older dog often needs more space from other dogs. He may avoid rough playmates and withdraw, or become more aggressive, in order to protect himself. As guardians of older dogs, it is important that we understand their changing needs and protect our seniors from unwanted contact.

We don't just have to stand by and watch them age; we can make the process smoother and less challenging.. Easy walks, specific exercises from the Tellington TTouch® method ground exercises, with easy obstacles taken slowly, along with physiotherapy, will help support the flexibility and functionality of the ageing body. The Tellington TTouches®, which are soft and mindful touches, applied all over the whole body to improve body awareness and release tension, are easy to learn and can applied for a few minutes every day.

A dog is only as old as he feels. Taschka (*see page 211*) still worked as a mould detection dog when she was 12, and there are many other examples. Katja's first dog, Sascha, was twelve when she passed her search and rescue dog certification; Baka was passionate about clicker training throughout her long life – she died at the age of 15.

Anatol, a Landseer male aged 11 (*lying down*), is still a fun playmate for the young Fredi, in spite of his age and physical limitations, which include arthritis in his left leg.

PART TWO:

COMMUNICATION SIGNALS

Although dogs have an oral language, it is not comparable with our verbal form of expression. They, therefore, specialised in a different type of communication. Besides the body language that we have discussed already, dogs also use subtle signals – predominantly facial expressions – to communicate. We will be looking at these in detail in this second part of the book. In most cases dogs use a combination of head and body positions to get their message across.

Generally, dogs use these signals to communicate with each other. Known as communication signals, they are mostly used to de-escalate a situation, to calm themselves or others, and to appease. Dogs also use the same signals towards people and other animals.

The signals dogs send us can be very clear, or they can be fleeting and elusive so there may be times when we barely perceive them. Depending on the dog's experience with other dogs, he will signal:

- Confidently and calmly
- With exaggeration, indicating insecurity
- Not at all, because he has never learned the communication signals or has been punished for using them.

This is comparable with the way people behave. Some are very polite and when they meet strangers they give them lots of space. In contrast, other people barge their way through a crowd and have no respect for personal space.

Body language is used by dogs in the same way we use it. We also look away when we don't want to connect with that person, walk off, or turn our body away to create more distance. However, dogs are far superior when it comes to reading body language, especially when it comes to the most subtle of signs. Their repertoire of 'small' signals is immense.

The ability to be aware, and correctly interpret, the tiniest movement of the eye, a change of facial expression, a breath, or muscular tension can be the deciding factor in whether a dog resorts to fear, aggression, or any other escalation of behaviour. Even the dog who apparently bites without warning has, in fact, already sent a mirage of subtle signals, which were sadly overlooked.

Since we have so many different breeds and types of dogs, some signals are unique or preferred by certain dogs. Therefore, it is crucially important for a dog to meet as many different dogs, and different types of dogs, as possible in the first few months of his life. These experiences should be positive because, as an adult, your dog will use them as references when meeting another dog of that type or breed. Because of this, it is advisable to work with an experienced dog trainer who can supervise a puppy play group. The puppies should play in groups matching their age, size and weight.

By learning to use their body language and communication signals early on, the dog will gain a solid base of communication skills which is immensely important for living in a world with lots of people, and lots of encounters with other dogs. Communication signals will not only be used by insecure dogs; confident dogs use them very successfully to calm those they meet, and clearly display that there is no need to worry. We have to remember that dogs always make the assumption that we understand their language. When under stress they will use more signals. Misunderstandings in communication happen when people either don't understand the signal, misinterpret it, ignore it, or, worst case scenario, punish the dog for using it. When the dog feels he is not understood, or receives the incorrect feedback to his communication from people, his trust and connection with that person will start to disintegrate. In the end the dog just resigns and no longer shows, or shares, his signals with that person. Inevitably, these dogs learn that they need to take care of their own safety, and frustration, insecurity, and aggression are the most likely outcomes.

These dogs are a high risk for people and other dogs, because they can no longer show their emotions politely. There can be a very smooth transition from growling to biting. However, it is much more dangerous when dogs have been punished for growling; those are the dogs who bite without warning.

We must be aware that dogs use their language to communicate with us. With good observation skills and a dose of 'listening to your gut feelings', you will find a way into your dog's emotional world, understand his needs, and be open to possibilities that could improve trust and change undesirable behaviour.

In the following pages, we have categorised communication signals on the basis of how frequently they are used.

1. TONGUE FLICK AND AIR-LICKING

REASONS FOR AIR-LICKING

Air-licking, or a tongue flick, is among the most frequently used signals in communication, and is often combined with a turn of the head. Air-licking can very be quick, where the dog darts out the very tip of his tongue, it can be a clear flick of the tongue touching the nose, or a deliberate, slow lick that includes the entire nose region.

Air-licking is not only shown with different intensity, but also in a lot of varied situations, such as on approach, when more space is needed, to request a break while playing, out of politeness, to self-calm or calm another, or when the emotions are undecided in a difficult situation.

To calm herself.

As a polite approach.

To calm and settle another dog.

Used by puppies, even when they don't yet have a lot of control over their bodies.

Air-licking is also used towards other animals. When dogs and cats live together, they both lick to communicate.

AIR-LICKING ON APPROACH

When approaching other dogs, especially strangers, air-licking often releases the tension, and helps to calm the situation.

Arwen also air-licks while she connects with the Dalmatian briefly before leaving with Ella.

With tension

Male Dalmatian

The liver spotted Dalmatian male has already met little Ella (*pictured*). In this case he is air-licking towards Arwen who is approaching him in the foreground. You can only just see Arwen, but she is approaching the male with a dropped head and direct eye contact (from his right). She feels responsible for her housemate, Ella.

- Clearly air-licking

Additional clues indicating tension:

- Ears raised at the base
- Wrinkles between the ears
- Slight whale eye towards Arwen, who is out of shot
- Tail moving

The Dalmatian wants to show through his air-licking that he has no bad intentions and he would like to calm Arwen.

Note Ella's confident posture when her big sister is close by.

In a situation with little or no tension, an air-lick is seen as a greeting and accepted as a polite gesture. The male Eurasian cross with the raised ears (*left*) approaches with a very relaxed posture. He air-licks as a form of communication.

Mild tension

- Air-licking

Additional clues indicating relaxation:

- Ears turned to the side
- Head barely above the level of the back
- Mouth slightly open
- Tail curled over the back (typical for the breed)

Now the Eurasian makes direct physical contact and raises a paw. He shifts his balance away from the other dog and respects his distance.

- Lifts paw
- Drops the head further

Respectful approach

The Golden Retriever, Amy, meets the very imposing Landseer male, Anatol. Her hunched posture and low wagging tail show respect, which is underlined by her raised right paw.

As the dogs get close and make direct eye contact, she licks her lips. The very confident male turns his head to show that he is neutral.

TO PREVENT AN ESCALATION IN BEHAVIOUR

The Elo hangs on his leash and accidentally floats above the Dachshund. The smaller dog flicks his tongue in the direction of the other dog to prevent a misunderstanding. Of course he may have also wanted to lick his lips, but looking at his tight leash, and the situation, it is more likely to be air-licking.

AIR-LICKING TO SELF-CALM

In these photos air-licking, or a tongue flick, is used less to de-escalate the situation, but more to calm the individual.

With great tension

Luke, the mature Border Collie (*pictured right in the photo sequence*) is not too impressed with the new family member, a four-month-old Border Collie called Jake. You can clearly see that he is not happy and does not know how to deal with the advances of the young and unruly puppy. The restricted space in their house makes it hard to move away, so he chooses to show severe threats towards the puppy.

In a larger area, Luke threatens a lot less and he can calm himself by licking his nose, and even Jake's nose. After a few weeks the two of them grew into a happy, bonded pair.

Accidental closeness

Arwen was very excited to see York, the Golden Retriever, and almost bumped into him accidentally. She ran very fast and headed straight for him. This kind of head on approach can easily lead to misunderstandings between dogs.

Arwen stops just in front of York and barely manages to turn without colliding. She air-licks to show him that she has peaceful intentions and did not want to provoke her friend. The air-licking in this case is very deliberate, a quick flick of the tongue would probably not have been clear enough.

When she takes off again, she licks again, reaching high up on to the bridge of her nose to calm herself.

INDECISION

Air-licking is used with many different emotions, ranging from joy to insecurity and from polite all the way to threatening. To judge the dog's emotional state when he is air-licking, we always need to consider the entire situation and the relationship between the dogs at that moment.

Accidental closeness, but friendly

Sally just greeted her neighbour Olli (*see Joy, page 457*) and would like to greet Quivive, aged 15, just as enthusiastically. Sally lies down for the greeting, but Quivive would like more distance. She may have already given the younger dog a quick warning, because Sally is looking at her and air-licking. The little brown dog wasn't sure how to deal with the situation because, in her excitement, she had missed Quivive's body language and signals, and accidentally got too close to her.

Air-licking combined with a threat

The grey female Berger, Verveine (*pictured right*), grew up with the black male Laguiole and left the family when she was four years old. The two dogs met again shortly after she had been in season. Laguiole showed a lot of interest in Verveine and pestered her. She clearly threatens him with her teeth. However, she has a lot of respect for the male so she reduces the impact of the threat by air-licking. Her exposed teeth show her arousal, while her air-licking shows her indecision as to how she should deal with the situation.

AIR-LICKING TO CALM

Sometimes puppies and youngsters are really frightened of adult dogs, and this far exceeds the respectful behaviour that would be considered desirable. A confident, older dog can teach the puppy to choose the correct communication signals, which will be helpful in easing his fears.

The confident older dog

The Tunisian wild dog puppy (*pictured right*) has her mouth wide open in fear. She barks loudly and shows her teeth. She is probably only this brave because her family members are close by. The adult male Husky mix air-licks to show her that all this crazy noise is unnecessary. He looks at the puppy briefly so he can then turn his head away.

Air-licking to request a break while playing

When it becomes clear that one of the wrestling dogs needs a break but the other does not respect the request, it is advisable to separate the dogs for a little while.

After a wild game with his littermate, the Rottweiler (*left*) air-licks to request a break

CONTACT WITH PEOPLE

A lot of dogs don't like to be touched. This is where a tongue flick or air-licking is used as a polite way of expressing discomfort. The dog will usually lean away and try to make himself appear smaller.

Small dogs are often manhandled and their individual sensitivity is easy to overlook. This is especially obvious when they are being picked up.

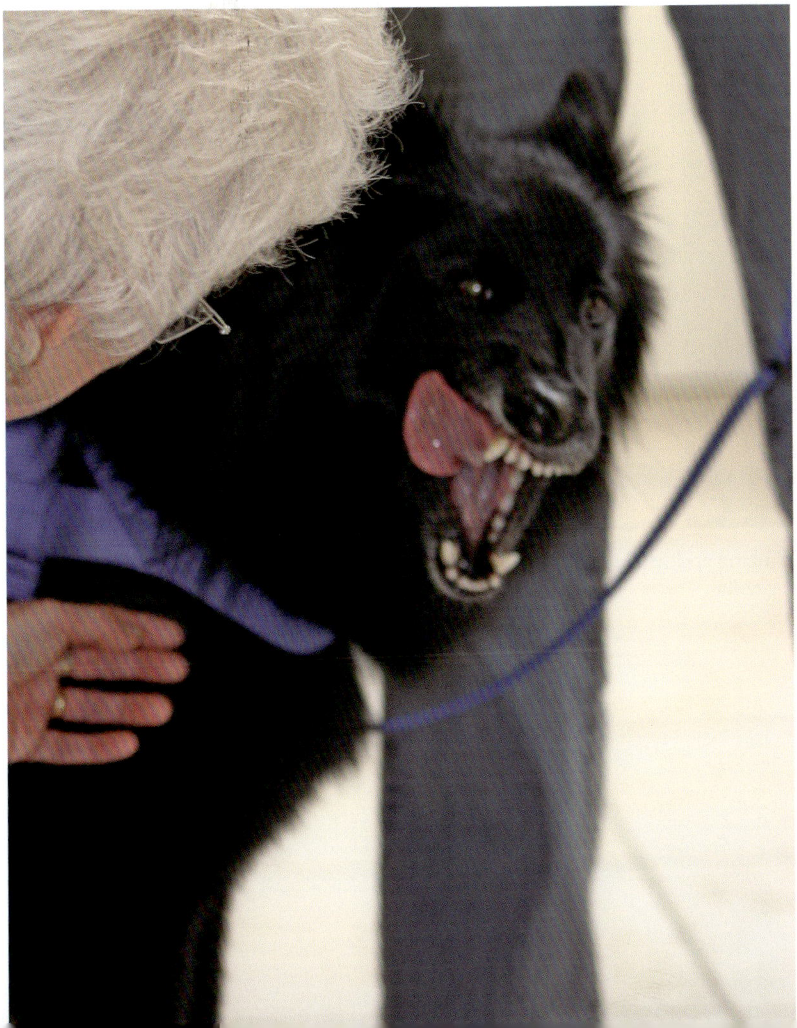

The need for distance

Kira, a female Groenendael, does not usually allow strangers to touch her. Here, Gabi works with an outstretched arm to respect the dog's need for space. She also touches the dog with the back of her fingers, which is less threatening. She looks in the direction of her hand to avoid direct eye contact with Kira. The dog can allow this touch, but her licking shows her disquiet.

Conflicted emotions

This dog lives in a shelter near Berlin, Germany. He had never been touched in a loving way until he arrived at the shelter. On the one hand, he is still very uncomfortable with someone being so close, and leaning over him. But, on the other hand, he has learned to enjoy the loving attention. Here he licks to show his conflicted emotions.

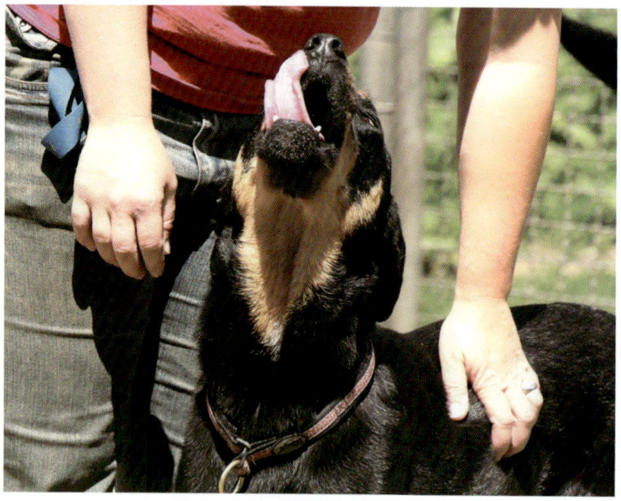

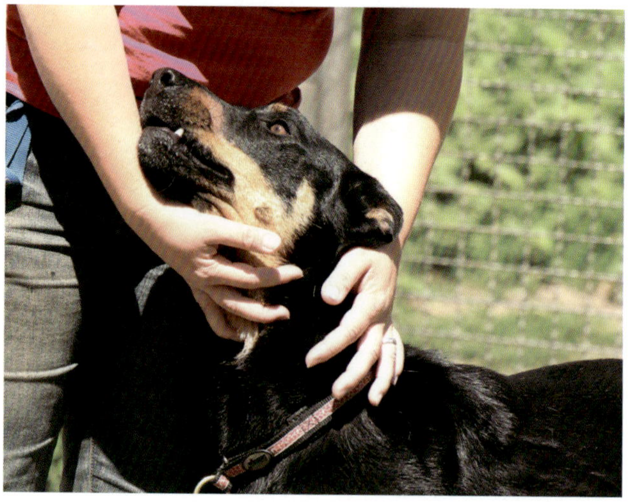

Pressure on the collar or leash

Outside pressure, such as someone grabbing the collar or harness, or pulling on the leash, is a very restrictive experience for many dogs. A tongue flick or air-licking is a common response to this unpleasant experience.

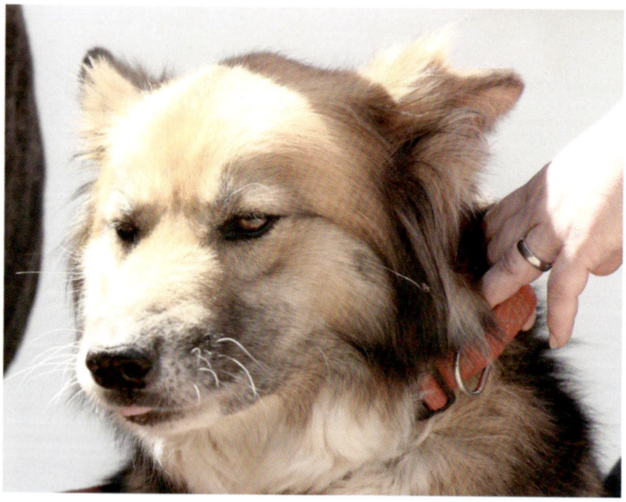

The pulled back ears and the expression in the eyes (blinking and whale eye), along with the tongue flick, shows how uncomfortable a dog is with pressure on his collar.

Mindful leash release

This sequence of photos shows a guardian taking great care when releasing the leash. We don't know if the Elo (pictured) had a particular problem with a hand being close to his collar, but it is always important to be mindful when clipping or unclipping the leash. Here the guardian reaches for the trigger clip from a distance to avoid bending over the dog. Despite this, he still feels uncomfortable and responds by air-licking.

Experiencing pain

Air-licking can also be a sign of pain, especially when it is used frequently.

Tiber accompanied us on walks until very close to the end of his life. While he would usually yawn at home to show his pain, he would air-lick repeatedly while we were walking.

With joy

Air-licking can be an expression of excitement and joy, although it is not universal.

- Licking
- Rounded back
- Tail low and wagging

Here Quivive is expressing her joy at meeting a favourite family member. She is so excited she air-licks to calm herself.

NON-VERBAL COMMUNICATION

These two dogs live in a shelter near Berlin. They know each other well and live together in a group.

The female, with the erect ears (*left*), is just having her harness put on. Therefore she cannot move away from the older male, Oskar, who is looking at her with direct eye contact. His facial expression is very clear, and it comes across more forcibly because of his facial markings. The female responds to his approach with a tongue flick and turns her head. Oskar notices the increased tension in the female and reacts by softening his eye and air-licking. He would like to de-escalate the situation.

2. BLINKING

REASONS FOR BLINKING

Like most communication signals, blinking is used in different situations and often in combination with other signals, such as turning the head, air-licking, crouching or lifting a paw. Blinking can vary from narrowing the eye to closing it; from a true, soft blink to a very tensed clamping shut of the eye. Blinking can also be shown asymmetrically, for example when a dog passes another dog.

REPETITIVE BLINKING TO BREAK DIRECT EYE CONTACT

Polite and confident dogs use this signal very often, but you also see it in those that are insecure. A blink is often immediately followed by a turn of the head.

This dog lives in a shelter and probably came from Eastern Europe. Every time he made direct eye contact with us, he immediately blinked. This is a typical and effective way for dogs to break eye contact and prevent fixating.

AS COMMUNICATION

Dogs will often blink to communicate with people, not just other dogs.

Blinking as a sign of affection: Banzai blinks at his guardian in a loving manner.

Blinking when meeting someone: this sloughi uses asymmetrical blinking as a dog passes him on his right.

Most street dogs are perfect communicators because they learnt the skills at a very young age, and their lives depended on it. These skills are more difficult to acquire if a dog spends most of his time on a leash, and meets more people than dogs. A former street dog, on a leash, may struggle with meeting other dogs because he is not used to being forced into close contact with people and with other dogs. The constraint, and lack of choice, is frightening to him. These dogs need our understanding and support if we take them into our family.

Relaxed in close proximity

These two dogs probably came from Eastern Europe and live in a group in a shelter. The light-coloured female (*pictured right*) blinks and turns her head away from the Greyhound, which is standing very close to her. She signals her peaceful intentions.

Relaxed and confident

Even when dogs get along well with each other, blinking can calm a situation even further.

The larger dog, Spot, is having a drink while the puppy, Robbie, approaches carefully. Spot raises his head and blinks towards him. The pup reads the signal correctly and comes closer in order to have a drink.

Spot feels completely confident in this situation. He turns his head away from Robbie to show him that he is welcome to come closer and enjoy a drink.

Relaxed communication

Blinking, whether with one eye or two, is a very commonly used signal in communication.

Here we can see some great communication in action. The brindle Sloughi would like to pass the light-coloured dog and wants to do so politely. Both dogs live together in a large family of Sloughis.

The brindle coloured Sloughi slightly arches his head and body while the light-coloured Sloughi blinks and turns his head so his friend has sufficient space to pass by.

IN PLAY

Tension due to being too close

It is easy to get too close while playing. In this case, the dogs involved use very quick communication signals to ensure the game stays friendly and doesn't escalate due to a misunderstanding.

These two Sloughis are having a great time playing with each other. In the midst of play, the younger dog (*dark collar*) ends up with both front legs on the older dog's back (*orange collar*). The older dog is not comfortable and shows this by exposing his teeth. You can only see the tips of the upper teeth, the gums are not exposed and the lower lip is hanging loosely.

The younger dog avoids eye contact with his playmate by blinking. He holds his ears flat back, and is keen to de-escalate the situation.

Growing tension while playing

These two dogs met at a dog beach and started to play. As the game progressed, the larger dog (*pictured right*) got a little too boisterous. The smaller dog blinks to request a slow down, or a break, in the game. He makes himself smaller and rounds his back.

Increased tension to avoid a provocation

Here we see blinking, with crouching, to ease the tension caused by lack of space.

Arwen (*pictured left*) makes herself really small, moves to the side, and blinks to make sure eye contact is not an option. Her ears are flat against her head and pulled back. You can see how uncomfortable she feels being so close to the white Swiss Shepherd, Darko.

HUMAN INTERACTIONS

We often cause dogs to blink when we get too close to them and fail to read their signals that tell us that we are in their space.

This dog is being cornered between the two children and does not have enough room to get away. The little girl is sitting right in front of him and holding his collar. The dog shows his tension through his pulled back, flattened ears, the tension in his mouth and the almost closed eyes. This is a typical picture of a dog who is very polite and puts up with the situation, but is clearly uncomfortable.

EXTERNAL FACTORS

Enhanced signal

In this case, blinking is enhanced by the dark pigmentation of the dog's eyelids and the white-haired "glasses" framing the eyes.

Hard to recognise

If a dog has long facial hair, it can be difficult to see whether he is blinking.

BLINKING FOR OTHER REASONS

Bright lights or eye irritations, as well as some ailments of the eye, can also cause dogs to blink.

Quivive is blinking because the sun was in her eyes when she heard her name being called.

If you take a closer look at the hair around this dog's eyes, you can see that it is crusty and the colour has changed. It is likely that his eyes run constantly and are itchy, or he may experience a burning sensation, which will cause him to blink.

3. HEAD TURNING

REASONS FOR TURNING THE HEAD

Turning the head is generally a very clear signal to indicate neutrality, disinterest, or a request for more space. This signal is often used in combination with others, such as blinking or a tongue flick and, in our experience, it is the communication signal that is most often used to maintain or increase space. The head turn can be obvious or subtle. Many reactive dogs have never learnt, or forgotten how to use, this signal when another dog upsets them.

Tension

When two dogs are in close proximity, and the situation becomes tense, the slightest possible head turn is expressed by turning the eyes aside and shortening the gaze.

We can see this clearly in the male dog, Rex (*standing on the right*), as Püppi walks past while looking directly at him. His eye is looking away from her, which you can see by the whale eye (*see inset*).

Relaxed

These dogs are turning their heads in a very subtle way. Neither dog changes their stance and the situation is relaxed.

The Jack Russell cross (*pictured right*) looks straight ahead as something has caught his interest. His ears are raised and pointing forward, and his tail is pointed upwards. The Dachshund joins the Jack Russell cross, but keeps his distance while he drinks.

The distance between the two dogs diminishes because the Jack Russell cross turns his head towards the Dachshund. At the same time, he lowers his head and his ears are more relaxed. The Dachshund responds by turning his head very slightly which, again, increases the distance between the pair.

RELAXED COMMUNICATION

Relaxed and disinterested

Independent, confident dogs will very often turn their heads away, for example, if they are disinterested.

Puppy Arwen really wants to make contact with the mature Golden Retriever, York, to show him how much she likes him.

York makes it clear to the puppy that he wants to be left alone just by using his body language. Besides turning his head, he leans away from the puppy.

Arwen has been a good student and uses the same body language, in this case to encourage the small Yorkshire Terrier puppy who is concerned about coming closer.

Relaxed and neutral

In this sequence of photos, the female Weimaraner (*left*) does not want to make contact with the male Rhodesian Ridgeback. The male would like to remain neutral and not provoke. Besides turning his head away, he also blinks, which breaks his eye contact.

Male Rhodesian Ridgeback

- Head turned away, no direct eye contact
- Ears held low
- Mouth closed
- Tail held below the level of his back

Male Rhodesian Ridgeback

- Head turned towards the other dog
- Head raised
- Tail carried higher

Male Rhodesian Ridgeback

- Head turned away again
- Eyes smaller
- Tail held slightly higher than back and moving

Male Rhodesian Ridgeback

- Turning the head even further
- Ears pointing forward more than before

Both dogs are now looking in opposite directions. They did not change the distance between them, or their posture, throughout the interaction.

From this angle, you can see how far apart the dogs were during this spell of signalling.

Relaxed and polite

When two dogs approach each other, you will often see that one of the dogs turns his head to show his disinterest in making contact or, if both dogs are moving, to signal his peaceful intentions.

Here, Pixel, the female Silken Windsprite (*pictured right*), shows her lack of interest in the approaching male by turning her head. The male confirms his neutrality by lowering his head and looking where he is going. His loose flopping ears, almond-shaped eyes, and slightly opened mouth demonstrate how relaxed he is. He shows a slight whale eye, away from Pixel. He has no intention of getting too close to her.

The male waits until he has gone past Pixel before announcing his interest. He turns his head without looking at her directly . His ears are slightly raised at the base in the flirting position (*see Ears, page 29*). His eyebrows are now slightly raised and his mouth almost closed, which shows his interest. Pixel now turns her head slightly towards him to let him know that she could be interested in meeting him.

Relaxed and polite

Trying to break eye contact

These photos were taken on a walk. The dogs don't know each other, and the Briard, Chocolat, does not know the area either. The grey dog turns his head away slightly and shifts his weight back noticeably. **Note:** in the second photo, the eye contact is broken.

TENSE COMMUNICATION

Tense and polite

To maintain distance

When walking along the beach, Quivive (*pictured left*) met a Great Pyrenean Mountain Dog, Taylor, with his guardian. The dogs were asked to sit side by side for a photo.

Quivive is not willing to get any closer to the imposing male; she clearly turns her head away and leans away from him. She is a little bit excited, which you can see by her long mouth. Taylor is the perfect gentleman; he turns away from her and blinks clearly to respect her need for space. His point of balance is shifted away for her as well.

Tension avoiding direct eye contact

Turning the head not only interrupts the eye contact, it also clearly says "this is too close for me."

The dog being carried has no other option, other than turning his head away to break direct eye contact, to tell the other dogs he needs more space.

Tension due to restraint and being too close

It is very common to see heads turned away in posed pictures.

This was an attempt to get Quivive and Tiber to pose together. Tiber moved a little closer to Quivive. As a result, Quivive wanted to leave but was prevented from doing so as Gabi's arm was encircling her shoulders. Quivive's only remaining choice was to turn her head away, and also lean away, to create more space and to signal to Tiber that he was too close.

Tension when people are too close

Both dogs avoid the hand that is reaching for them by turning their heads away. Some dogs will show this response only when strangers reach for them, but there are also dogs who will try to avoid their guardians.

In addition to turning his head, this dog is also clearly leaning away.

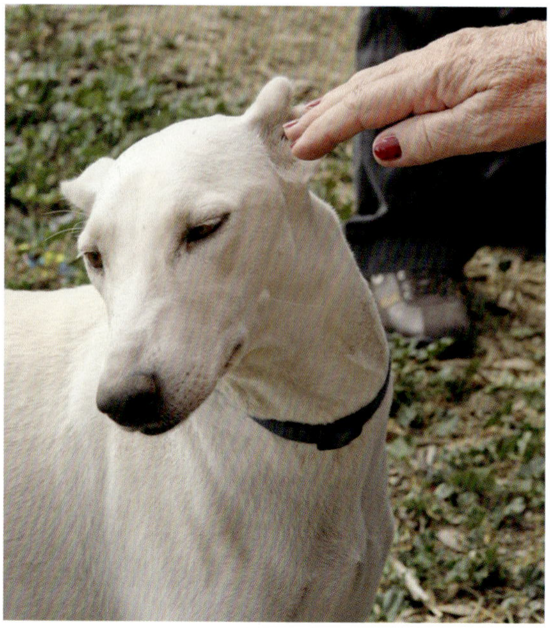

Here the dog also blinks while turning his head away.

TO CLARIFY THE INTENTION TO PLAY

Turning the head away while playing confirms a peaceful intent, even though the game looks rough.

Tamika (*left*) turns her head away from Arwen with her mouth wide open to show the younger dog that they are still having fun.

This is also a fairly common situation when dogs are playing. Spot stands over Rani who is lying on the ground. By turning his head, he makes sure his position is not misinterpreted.

COMMUNICATION AT CLOSE QUARTERS

In the two following photo sequences, you can see that when the black-and-tan male turns his head, it is not a random movement; it is a conscious signal. Both dogs continuously communicate with each other as the situation develops.

Both dogs are moving towards each other head on. The black-and-tan male is at a trot. Since there is very little room for him to move out of the way, he clearly turns his head away from the female in the foreground. This supports his friendly intentions.

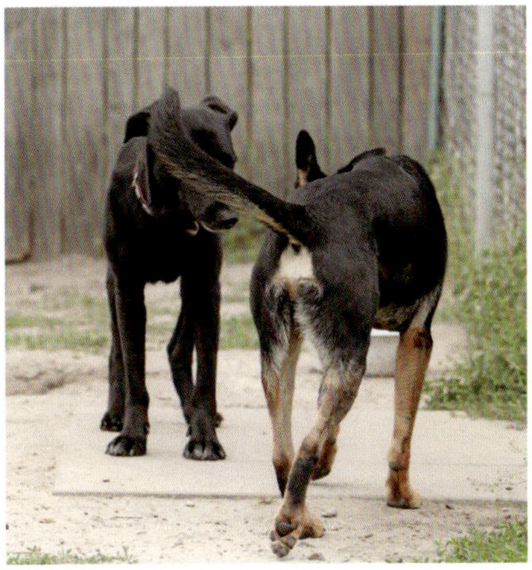

The black-and-tan male walks up to the black female. He keeps his head low, his tail is level with his back and moving. As he gets closer, he lowers his head even more and turns it away. He walks past the female without looking at her. His ears are held towards the back, the level of his head and tail remain unchanged. The black female keeps her head level with her back, her ears are pointing towards the back. We cannot see her tail. She blinks and stands motionless.

The black female changes her ear position only after the male has passed. She now points them forward and doesn't hold them as close to her body. The female stays in the same spot during the whole photo sequence.

4. TURNING THE BODY AWAY

REASONS FOR TURNING THE BODY AWAY

Turning the body, which is often followed by walking away and leaving the situation, is used to create more distance, often when turning the head away is insufficient.

It is usually preceded by leaning away, shifting the weight, or turning the head. Turning the body away is either a decision to influence what happens next, or the response to a request from another dog to give more space. It is also used to show disinterest, as a response to a stop signal from another dog or person, or as a polite approach. If both dogs turn away as they approach each other, they usually end up passing in an arc and remain neutral with no contact.

OUT OF RESPECT

Champs, a male Yorkshire Terrier would like to play with the female Bulldog, Lola.

Showing interest

Yorkshire Terrier, Champs

- Body turned towards her
- Ears in flirting position
- Tail held lower than the back and wagging eagerly

As Lola turns towards Champs, he turns away. The direct eye contact from Lola, possibly exaggerated by her wrinkles, is a bit too much for him.

To protect distance

Yorkshire Terrier, Champs

- Body turned away
- Ears pulled towards the back
- Tail unchanged
- Whale eye

Marley, a Poodle, is interested in Edward, an older Spaniel, However, Marley's attention is too much for Edward and he turned away – maybe to avoid potential confrontations.

Here Edward clearly makes contact with Marley. Marley's ears are slightly pulled back. You can tell by the almost raised left front foot that he is shifting his balance away from the much younger dog. His mouth is nearly closed.

This seems to be too close for Marley, possibly because the Yorkshire terrier, Champs, who is hidden in the previous photo, is also standing really close to him. Marley's mouth is now a little more open.

Marley turns away from Edward and walks off. His mouth is clearly open which gives the impression that he is more relaxed.

POLITE COMMUNICATIONS

This Sloughi family lives in Tunisia in their breeder's yard.

The white livestock guardian dog, on the very left in the photos, is also part of their pack. At this moment there are a lot of communications between the different dogs in order to ease tension around the gate. Both the guardian dog and the brindle Sloughi are clearly turning away their bodies.

The guardian dog moves away from the gate at the same time as the brindle coloured Sloughi, and there is some unwanted closeness with eye contact.

Both dogs avoid frontal eye contact by turning their heads. As they are walking forward they are also starting to turn their bodies away from each other. Both of them keep their heads clearly turned away from each other, and any other dog in the group. As they walk away you can see how, as their heads turn away, their bodies follow the same direction. The behaviour shows neutrality to avoid unnecessary provocation between the members of the family.

5. WALKING IN AN ARC

REASONS FOR WALKING IN AN ARC

Walking in an arc increases the distance from another dog, a person or an object in a sideways direction. Dogs use this behaviour frequently, mostly when they want to stay neutral, to be polite, or to show respect. A dog may also use this form of communication if he feels crowded or threatened, or when another dog or person has come too close. Walking in an arc is a means of de-escalating a situation.

TO BE POLITE

Even the smallest arc is appreciated, as can be seen in this polite meeting between two Sloughies, who bend their spines in a convex shape. They maintained this same distance throughout their encounter, and both enjoyed the stress-free experience .

Both dogs

- Convex bend to the spine
- Tail (brindle coloured Sloughi) level with his back and gently wagging
- Heads stretched forward

REQUEST FOR SPACE

You can clearly see that the male Golden Retriever, who is standing still, is asking for more space. He would like to remain neutral and trusts that his posture and presence will not be misunderstood.

Rani (*foreground*) approaches the male, who stops and remains motionless with his head held high. He looks over the top of her with no eye contact. His tail is level with his back and moving.

Rani is still approaching the motionless male. Her tail swings level with her back, her ears are flat against her head. She would like to make contact with the male, but hesitates because of his posture.

As Rani approaches him head on, he turn his head and creates eye contact. His posture remains unchanged. Rani understands and starts to walk in an arc. She starts by first turning her head, and then her body, away from the male. She then drops her head. The male remains motionless until she has passed him, in a large arch, and disappears into the forest.

TO AVOID

Walking in an arc is used to create more distance to the sides, and to avoid conflict.

In the following sequence of photos, you can see Deluxe, a female Weimaraner (*pictured right*) executing a nicely-curved arc as the reddish-brown male turns towards her.

Note the Weimaraner's behaviour:

- Point of balance away from the other dog
- Tail carried higher than the back and moving
- Ears hanging down loosely

The male makes eye contact and turns towards her. The Weimaraner's avoidance is now more obvious. She is holding her tail higher and tighter and, very subtly, turns her head towards the male. He takes note of her behaviour and slows down.

The male turns away from the Weimaraner and breaks eye contact. Now she look at him. Her point of balance is almost back to neutral.

The male's polite response encourages the Weimaraner to make contact with him.

TO REMAIN NEUTRAL

When approaching a group

When approaching a group of dogs that are very interested in each other, there can be an escalation when the meeting is head on and /or happens too quickly.

In this sequence of photos, the black female, Taschka (*standing in the middle*), is being sniffed with great intensity by a male (*in the foreground*) who is a stranger to her. The Australian Shepherd cross, Spot, who spends a lot of time with Taschka, is approaching from behind. He wants to find out if his friend is in danger.

Spot makes eye contact with Taschka. He does not approach the other two dogs straight on, but does so in an arc.

As Spot passes Taschka, the male comes to the other side and lifts his head. He shows a little bit of a whale eye (*see inset*) his mouth is long and his ears are flat on his head.

Spot widens his arc and therefore his distance to the male. Spot's ears are now towards the back, his lips are pulled back. He only brings his ears forward again after he has passed the two dogs. He looks into the direction he moves, and the male watches him with soft eyes.

RELAXED TOGETHER

This is an everyday scene where two dogs communicate with each other and a third (the Greyhound) passes in a small arc so he does not disturb them.

These dogs live together in a shelter where they are mostly outside in large groups. Here, the older dog (*in front*) moves a little too fast. However, the other dog is approaching in a very relaxed manner, which can be seen by the position of his ears.

Both dogs

- Tails held high
- Direct eye contact

Male mixed breed (*right*)

- Ears loose and held to the side
- Eyes small
- Mouth slightly open

While the two dogs make contact and communicate, the Greyhound (who also lives with them) approaches from behind. He immediately uses an arc to pass them. He seems tense, which we can see by his open mouth, his pulled-back ears, his rounded spine and low head carriage. However, the impression is false. It was a very hot day and he is panting due to the heat. His posture is very typical for the breed. He is actually passing the two dogs in a quiet, relaxed manner.

TO AVOID CONFLICT

This is another group interaction, photographed at the shelter.

There is some tension between the two dogs. They are fixating on each other, although the body of the Shepherd cross (*front on*) is already hinting at a slight arc.

Shepherd cross
- Ears pointed forward
- Direct eye contact
- Mouth closed

Both dogs
- Tails raised high
- Heads lowered and pushed forward

As the grey-white male approaches, the Shepherd cross clearly arcs to avoid contact. He looks ahead and does not respond to the other dog's eye contact. His tail is now a little lower and his ears turned to the side. These are clear signals aimed to de-escalate the meeting with the white-grey male.

FOR SAFETY

An arc can also be used as a safety measure. The distance the dog creates will vary from situation to situation.

This Tunisian street dog is carrying some food in his mouth when he catches sight of Katja. You can clearly see by the centre line on the road that he uses an arc, which is just large enough to keep himself and his food safe, while getting round her. He speeds up slightly as he passes Katja and then continues on his way.

SPACE MATTERS

Creating and maintaining distance plays a much more important role in a dog's life than we realise. Most communication signals are for the purpose of creating and maintaining a safe space.

Walking in an arc is most commonly used to stay at a safe distance. In the case of meeting another dog, the extent of the distance depends on the individual dog's confidence and experience, as well as the other dog's signals, or lack thereof.

It is very important for a dog to learn through experiences that he can feel safe and secure when he creates distance. This is especially applicable to dogs who are leash aggressive. The aggression usually starts when the dog feels insecure about being too close to other dogs. The leash, or any other physical restriction (such as a doorway or a narrow path) limits his ability to create space so the dog feels he has no option other than to show aggression. From his perspective, approaching another dog face-to-face (which often happens when dogs are walking on-leash) is both impolite and a cause for concern. He will, therefore, react accordingly. This is most particularly the case if a young dog, who lacks the experience to read communication signals, lunges at another dog front-on. This is very likely to trigger an aggressive response in a reactive dog.

When exercising your dog on-leash, always walk in an arc when meeting other dogs, even if he does not have any issues with heading straight towards other dogs. Praise him to encourage this behaviour, and teach him to be polite. If your dog is insecure about meeting other dogs, you can help him by walking in a large arc, moving as far to the side as the situation requires. Sometimes that means you have to cross to the other side of the street, or turn around and walk in the other direction.

If your dog is unable to take a treat politely because he is too close to another dog, he is telling you that he needs more space. The ability to take food calmly, without grabbing it, is a good indicator of how much space a dog needs.

TO EVADE A PERSON

An arc can be used to move away from a person. If a dog is afraid of a person, and restrained by a leash, he does not have the option to create a safe distance. You can gain a lot of trust and respect from your dog by being mindful of his need for space and walking in a large arc with him when he needs it.

Golden Retriever, Rani, is just coming out of the lake and the guardian of the chocolate Labrador talks to her. Even though she usually likes strangers and always greets them happily, she walks towards the dog and looks at the man with a whale eye.

Rani makes contact with the Labrador, but also keeps an eye on the man. As he bends forward and talks to her again, Rani shows very clear signs of avoidance and jumps to the side.

Rani chooses to make a large arc around the man who is clearly causing her concern.

IN UNCERTAIN TIMES

Here we can see how quickly external influences can make communication between dogs more difficult. Even dogs with good eyesight may feel insecure when exposed to poor light conditions.

Here, communication is made very challenging due to the dense fog. Dhanyi and the smaller dog, Ella, are in the foreground and her housemate, Jean, is some distance ahead, facing them, and a little hesitant to approach. Jean and Dhanyi are looking at each other; Jean is not really sure if Dhanyi recognises her as part of the family.

Jean turns her head away from Dhanyi, which clearly shows that she does not want to approach.

Dhanyi turns her head and looks at Katja for advice. This eye contact, which is a form of checking on, should be seen by guardians as an opportunity to recall the dog, thus removing him from a difficult situation.

Jean decides to approach slowly in a large arc. Dhanyi watches her, but stays motionless.

Since Jean stays on the arc, creating sufficient distance, Dhanyi can turn her head away.

Dhanyi has now turned away completely and happily runs towards her guardian. Jean keeps walking and sniffs the ground. She is now much more relaxed.

6. SLOWING DOWN

REASONS FOR SLOWER MOVEMENT

Slowing down can often be observed when dogs meet. A dog may run full tilt towards another dog and then, at a certain distance, start to reduce his speed. This is either a polite gesture between dogs, or one of the dogs has sent a signal asking the other to reduce his speed and not come any closer. When the request is answered correctly, the dogs will usually meet each other in a friendly manner, accompanied with slow movements and a sideways approach. They will generally employ other communication signals at the same time, such as blinking, turning the head away and air-licking.

Rising tension between dogs can often be seen when they increasingly slow down their movements. The more tense the situation, the slower the movement becomes, leading all the way to stillness. We usually see very stiff movement with braced joints, a very high tail that either does not move at all, or moves very fast at the tip. A young dog that misreads these signals when approaching an older dog often fails to slow down, and is then reprimanded by the older dog. The guardian of this type of dog is often surprised that their friendly dog keeps getting attacked, or even bitten, when he just wants to play. However, it is his inability to slow down that is causing the problem.

People also dislike being approached at high speed. Dogs will usually slow down when approaching people in order to appease. When we call a dog to us, using a strict tone of voice and appearing angry, he may come quickly at first, but as he gets closer he will slow down. Our mood and posture tell him that it is better not to come too close; it will be safer to keep his distance.

WHEN APPROACHING OTHER DOGS

Some dogs have been bred to approach quickly, such as the Australian Cattle Dog, who needs to use this tactic when herding cattle. However, this seemingly uninhibited behaviour can lead to misunderstandings with other dogs. Therefore, it is very important that your dog can be recalled at any time. When playing ball it is important to set up the game so dogs don't meet front on.

This Jack Russell Terrier mix is running towards a group of dogs at a gallop. His ears are pointing forward, his tail is held a little higher than the level of his back and it is moving. He looks very bold and is carrying a ball in his mouth.

As he sees a group of six dogs, he slows his speed and stops. He shifts his point of balance towards the back. His tail and ears stay upright and he watches the dogs, which are now approaching him.

WHEN NOTICING ANOTHER DOG

Slowing down will also be used when a dog is unsure of a situation and feels it is safer to approach with caution.

The Dalmatian is approaching at a brisk trot, looking straight ahead.

- Tail raised high with little movement
- Neck raised
- Head held mostly forward

As he changes the direction of his glance towards the camera, he stiffens his tail and slightly lowers his head.

The Dalmatian is now sideways to the camera. He slows down and comes to a stop.

- Tail held high and moving
- Neck raised

- Head stretched forward and raised

This is not a relaxed stop; the Dalmatian is very stiff in his posture. It seems he has seen something that causes him to hesitate.

FOLLOWING A CONFRONTATIONAL APPROACH

This is the large group of dogs that live in Tunisia. They are kept together in a large space and only leave the farm to hunt. Here, the light brindle-coloured female Sloughi leads a charge towards a stranger. Most of them were quite friendly and interested, but this young female was clearly alarmed.

The young female (middle) barks at Katja as she approaches. You can tell by the backwards shifting of her weight that she is already slowing her speed.

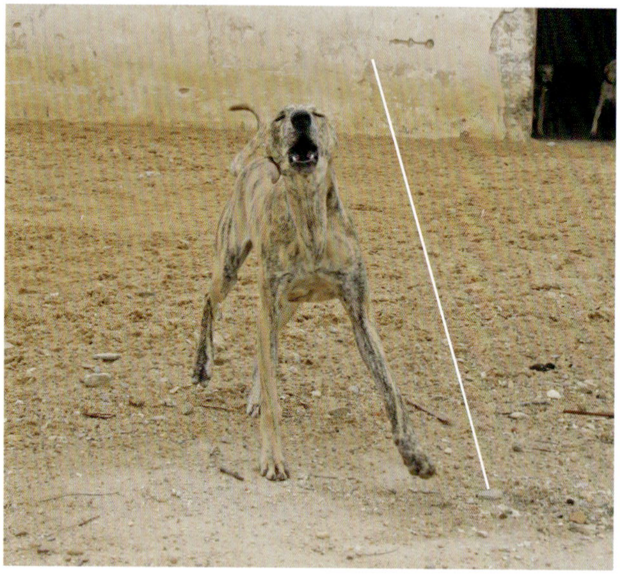

She reduces her speed even more and shifts her weight far back.

Now she is just walking, which gives her a better chance to assess the situation and decide what to do next.

FOLLOWING A BOISTEROUS APPROACH

Young dogs find it especially difficult to approach slowly.

Rani (*standing on the left*) meets a young female Golden Retriever who approaches very quickly. Rani remains standing still; she is holding her head high as well as her tail.

When she sees that Rani is motionless, the precocious youngster slows her speed. The shape of her mouth has changed and she has raised the hairs on her back. She is now a little uncertain as to whether she should continue her approach.

Rani remains motionless and has probably created eye contact – only her tail moves. The young dog slows her speed the closer she gets.

The young dog's mouth is now closed and she is showing wrinkles on her face. She has dropped her tail, and her ears are pinned close to her head and back. While Rani has shifted her point of balance to the front, the young dog has shifted hers to the back.

TO AVOID A SITUATION

In the foreground the smaller, black dog, Taschka, and the tri-coloured male are very interested in one another. A Jack Russell Terrier approaches at a trot behind them.

- Head carried high
- Tail raised high
- Ears pointing forward

As soon as he notices the two dogs, who are blocking his path, he slows down.

As he gets close to the two dogs, he turns away.

- Head slightly lowered
- Tail level with the back
- Ears held towards the back

His posture has really changed. It is highly likely that Taschka, who is facing him, made eye contact that said. "Stay away! Don't disturb me!"

He turns to run back and finds himself facing Spot, who just appeared in the top left of the picture. The Jack Russell rounds his back and drops his head and tail. He is no longer comfortable in this situation.

7. CROUCHING

REASONS FOR CROUCHING

Typically crouching, or making oneself smaller, is described as submissive. The next level of this crouching behaviour is lying down, which is generally perceived as a means of averting an attack. However, in our experience this is not always the case. This posture is often adopted when a dog is feeling insecure; his legs are angled and his head and back lowered.

Many dogs approach a loved one in this manner, and we see it as a way to convey a sense of belonging. It is also a form of polite greeting so, once again, we need to look at the entire situation rather than relying on a single communication signal.

Some dogs crouch easily, others rarely use it unless they are in a very threatening situation. It greatly depends on the individual dog's character, but can also be influenced by his anatomy or physical well-being. An older dog may experience pain if he attempts to crouch. In contrast, young dogs over-use this signal, often rolling on to their backs when meeting older dogs they know well.

AS A SIGN OF RESPECT

Towards an older dog

Crouching may be used as a sign of respect towards an older, familiar dog, but it can be seen as pushy and may annoy the other dog.

Arwen (*left*) approaches the older dog, Ronnie. She has known Ronnie since she was a puppy and wants to come closer. She shows her respect by crouching and carrying her tail low while wagging it.

A respectful friendly approach towards a familiar dog

If the mature dog shows that he is annoyed, the puppy or young dog should be called back immediately using a happy, positive tone of voice.

Arwen approaches the mature Golden Retriever, York, with a crouching posture (puppy advance). She makes herself small and flicks her tongue in order to not upset the older dog. Her point of balance is far towards the back. York does not return Arwen's eye contact. By leaning away and pulling his ears back, he shows the puppy what he thinks about her approach. He wants to be left alone.

Respectful approach (Arwen)

- Severe angulation of the legs front and back
- Severe bending of the elbows
- Tail held low
- Back rounded and lower towards the front
- Point of balance shifted to the back
- Head stretched forward
- Ears hanging to the side and back
- Tongue flick

Many dog guardians still believe that puppies, and youngsters, receive some sort of special protection when they are meeting strange dogs for the first time. Sadly that is not the case. The puppies' mother is usually very tolerant and gentle with her offspring as long as she lives with them. Other dogs in the family can – as long as they have been with the puppies since birth – behave in a similar way towards 'their' puppies. However, there are also mothers who are very rough and inappropriate with their puppies.

This patient behaviour is rarely seen when adult dogs meet young puppies for the first time. Their behaviour towards the young dogs greatly depends on their previous experiences. Many adult dogs will snap at puppies, which may lead to injury.

TO APPEASE AN ADULT DOG

Many dogs who are young, or still feel young, will crouch or even cower.

In these photos, the clumsily approaching young male Labrador (*on the right*) triggered the mature female Weimaraner's intolerant response. He stops when the female shows him clearly that he is not to come any closer. He shifts his point of balance to the back and makes himself small. He pulls his ears towards the back; his tail is low and wagging.

The young Labrador would like to appease the Weimaraner and tries to reach for the corner of her mouth, so he can lick it. Clearly, this is too much for her. She raises her eyebrows and stares. Her ears are pulled back and you can see wrinkles around her nose. She avoids the young dog by leaning back. Her tail is low and she is very tense.

We noticed long threads of saliva were coming from the female's mouth and her gait was very stiff with a tight, raised back. We assume she may not have been feeling well, which could be why she objects to the young male coming close to her.

INSECURE IN A NEW SITUATION

Rani, the Golden Retriever, is a little alarmed by the brightly coloured model cow in downtown Berlin. Katja sprinkles some treats to show her there is no cause for concern.

Feeling insecure

- Crouched posture
- Severe angulation of the legs
- Point of balance towards the front
- Topline falling towards the front
- Tail close to the body
- Head and neck stretched forward

Rani slowly approaches the model cow looking for treats, with an extended neck – that is why her point of balance is so much over her front legs. The crouching posture indicates her insecurity; she is ready to flee if the cow decides to attack. Tellington TTouches® could reduce her fear in this situation.

Here, Rani is meeting horses for the first time. She is concerned, but she is also sufficiently curious to venture close to them.

Feeling insecure

- Crouched posture
- Back severely rounded
- Tail close to the body
- Head carried low
- Eyes rounded
- Ears held towards the back

WHEN IN TROUBLE

Our posture is influential and can cause a dog to cower. Needless to say, we strongly disapprove of this type of reprimand.

This dog ran off and didn't listen right away when his guardian called him back. As a result, he receives a major telling-off. You can see his front end ducking away from her and he is leaning to the side as much as the very tight leash allows.

As if the telling-off was not enough, the dog was then hit with a towel. He probably knew that the punishment was coming so he cowered, and pulled back, in anticipation. This behaviour is often misinterpreted as "he knows exactly what he did wrong, and why he is being punished." In reality, this dog is not suffering from a bad conscience, he is pulling to the side due to the opposition reflex that is created by the tight leash (*see The opposition reflex, page 177*).

IN FEAR

When dogs are off leash, a boisterous, undisciplined approach can cause real fear, even though no physical injury is inflicted.

When out on a walk, Rani is suddenly surprised by a German Shepherd speeding towards her. She bends her front legs to make herself small, and is ready to get out of the way if needed. Her ears are pulled back, her mouth is closed and her eyes are rounded. She is very tense. The German Shepherd cannot stop, but his point of balance is clearly away from Rani.

Rani rounds her back and ducks away from the Shepherd.. Her eyes are very round, and her tail is pulled far under her belly. This situation has scared her.

WHEN ANOTHER DOG IS TOO INTRUSIVE

The young female Eurasian, Diana (*left*), keeps pestering Quivive to play. The more mature Quivive is not interested and tries to keep walking, as she ignores Diana's over-the-top behaviour. However, Quivive is not as strong as Diana, and so she is concerned by the situation. She has probably lifted her paw to appease the stronger dog.

In this case, Diana's youth, size and energy play a role in Quivive's decision not to respond to the request to play. Young dogs are often rough and boisterous, which makes it more difficult for an older dog to influence and guide the play. In this type of situation, the dog's guardians should step in and help. In this case, Diana was put back on the leash to give Quivive a break.

Quivive (*right*)

- Crouched posture
- Back very round
- Tail low and carried close to the body
- Head carried low
- Ears pulled back
- Left front paw raised

AS A REQUEST FOR A BREAK

Rani (*left*) and Spot have had a fun, exciting wrestle but now Rani is asking for a break. It may be that the game became a little too much for her.

- Back very rounded
- Head low
- Tail carried low and moving
- Legs bent

Respect for another (larger) dog

Golden Retriever, Amy, meets the very large Landseer, Anatol, at Gabi's training venue. She is very impressed by the big male dog and keeps a cautions distance.

Amy approaches with her head dropped and tail low. She wags just the tip of her tail. She is pulling her lips back so her mouth is long; her ears are flat and pulled back. She is very insecure and does not want to provoke Anatol.

After a very careful approach from the side with direct eye contact, Amy dares to step in front of Anatol. She still cowers, but Anatol exudes calmness and neutrality.

INHERITED INSECURITY

Cookie is a very insecure dog who has a lot of fears, especially when she is outside. She started crouching shortly after she was born and continues to exhibit this type of behaviour.

When insecure

- Crouched posture
- Back very rounded
- Point of balance shifted to the back
- Tail carried low and close to the body
- Neck slightly raised
- Head dropped and stretched forward
- Ears held towards the back
- Rounded eye

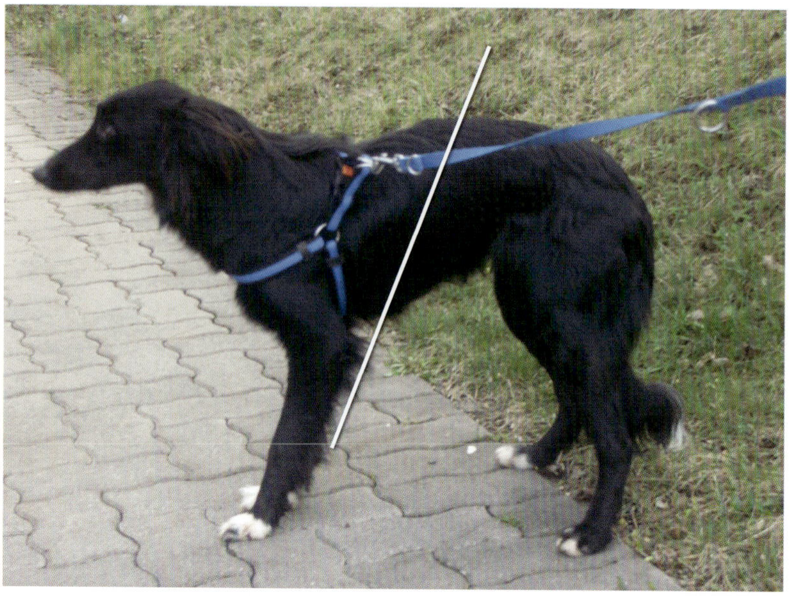

Cookie is wearing a Tellington Body Wrap which is often used in the Tellington method to improve the dog's awareness of its body.

With the Tellington Body Wrap

- Back less rounded
- Tail raised and moving
- Neck and head raised
- Ears towards the front

Sometimes a posture can be adopted out of habit and then influences the psyche or the emotional mood. The Tellington Body Wraps give Cookie's body different proprioceptional information and allow her to let go of her habitual posture, which does not serve her well.

HUMAN APPROACHES

We can easily trigger a dog to crouch, or even cower. This most often happens when we reach for the dog's head from above. Many dogs, even if they don't have any problems with being touched, dislike this type of approach.

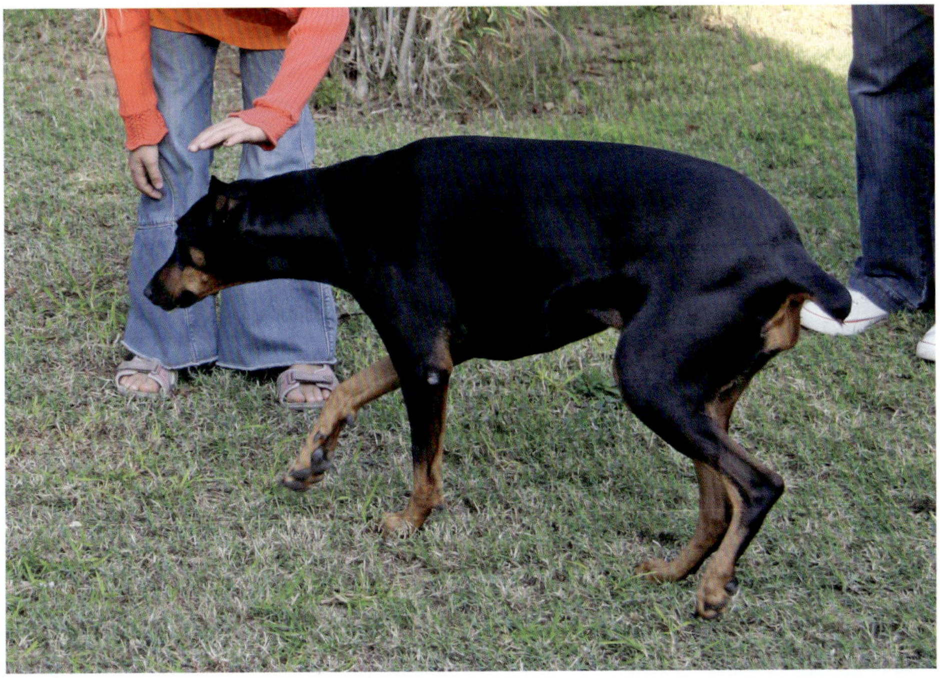

Crouching, combined with a dropped head, low tail and pulled back ears, is often seen when a hand comes down over a dog's head. You will also often see other signals, such as blinking, lifting a paw and turning the head or body.

WHEN PEOPLE ARE TOO CLOSE

This dog clearly shows signs of stress, which is made worse by the hot weather. He is badly in need of a break and should be taken to a quiet area where he will be able to calm down.

This Golden Retriever finds himself in a crowd of people and feels very uncomfortable.

- Hind-end sunk all the way to the ground
- Crouched posture
- Tail between the legs
- Ears flat against the head and pulled back
- Top of the head tight
- Eyebrows raised
- Mouth long
- Long tongue and edges of the tip are rolled up

8. SITTING

REASONS FOR SITTING

Sitting, and lying down, are strong signals that are may be used in distressing situations. However, both signals are employed to varying degrees. Some dogs never make use of them; other dogs, or specific breeds, favour these signals, even in less threatening situations. In very threatening situations, such dogs have a tendency to 'freeze'. In addition to appeasement, these signals serve as a request for distance; they are used by confident dogs to demonstrate that their approach is harmless.

A TENDENCY TO FREEZE

Rani, the Golden Retriever, is a newcomer to this puppy play group and she is being examined by a number of puppies at the same time.

With tension

- Sitting down

Additional clues indicating tension:

- Centre of balance shifted towards the back
- Ears pulled back
- Mouth closed
- Vibrissae raised
- Whale eye

Rani sits down to cope with the many puppies that surround her. She is not comfortable in this situation. The puppy in front of her is too close and makes direct eye contact. Rani tries to break this by turning her eye away and showing a whale eye. The puppies beside her, and behind her, are also too close. Rani feels threatened; she does not dare to get up and walk away. She appears to be frozen.

This is the type of situation where the dog who feels threatened will often freeze to prevent an attack from the surrounding dogs. As the dog's guardian, we need to deal with this situation very carefully – particularly if adult dogs are involved – and not blame our dog for being 'disobedient' if he doesn't come when called. He is only allowed to move when the other dogs give him permission. It would be helpful to encourage the other dogs to turn away and leave.

TO APPEASE ANOTHER DOG

Chemsi, the small, female Sloughi, has just arrived in her new home, and is getting to know her future partner, Inca, a female Dobermann.

Chemsi grew up in a breeding facility with a lot of adult dogs and, in this situation, she is the confident dog. Inca feels insecure and also irritated by the puppy. You can see her excitement by the drool in the corner of her mouth. Chemsi tries a lot of different signals, at the same time, to calm Inca.

To appease (Chemsi)

- Sitting
- Turning the body away
- Turning the head
- Blinking

REQUESTING A BREAK

This is a scene Katja observed at a doggy day-care in Dubai. The black, mixed breed male keeps inviting the small, female Cocker Spaniel to play. However, when they played, the game was often too rough for her, so she would sit down to stop the play. As you can see by the hands on the young black male, he did not understand or respond to her request so the caregiver had to gently interrupt the interaction to give the Spaniel a break.

The female Cocker pulls her ears back and her eyes are wide open. She has also pulled the corners of her mouth back and her mouth is slightly opened. Her head is raised high as she tries to maintain eye contact with the male. She is not afraid, but she needs a break, which she hopes to receive by sitting down. (*see Splitting, page 477*)

REQUESTING SPACE

Sitting down is often used to request more distance.

The male Golden Retriever (*left*) is very interested in the female. He maintains a certain distance, and his point of balance is shifted away from her, but he is intrusive enough to push her tail to the side. The female is not comfortable. Her back is rounded and she keeps her head level with her back. She lifts a paw, her ears are pointing back and she has shortened her lips.

The female Golden Retriever sits down, remains still, and directs her gaze elsewhere, to make it clear to the male that she is not interested. It could be that she has recently been in season, or is just about to come on heat, which would explain the male's particular interest. He responds to her request and moves away. You can tell by her unchanged posture that she is still very tense; only her ears are starting to relax and point forward.

This type of behaviour is often seen in female dogs when pursued by a pushy male. The female in question has no interest in the interaction and will often sit to protect her genital area. In this instance, the person responsible for the male needs to recall their dog and put him on the leash, if necessary. Males with a high sex drive will rarely choose to cease their unwanted attentions. If you are the guardian of a female dog, you should be prepared to protect your dog from advances of this sort. It is also important to be aware that weeks before a female dog is in season, she releases hormone induced pheromones that are irresistible to males.

WHEN INSECURE

Or wanting to appease

Micky, a Terrier mix, comes in too fast as he approaches Daniel, a male Golden Retriever. Micky is a fairly insecure dog when it comes to meeting other dogs and tries to make up for it by being too assertive (*see The Body, page 210*)

Micky is always far too pushy and, here, he climbs up on Daniel. The male Retriever is a very kind dog and responds confidently, showing a fair degree of relaxation. His tail is level with his back and moving.

Daniel shifts his balance forward, just for a moment, and Micky sits down immediately, and shifts his weight away from the larger dog.

When Daniel moves, Micky adds licking and lifting a paw to his communication signals.

9. LYING DOWN

REASONS FOR LYING DOWN

Lying down is a very clear signal. It is used, as we have already mentioned, as a preferred method of communication by some individuals and is more common in certain breeds. Some dogs never make use of this signal.

PROTECTION MEASURES

If your dog is lying down, and there is no imminent danger, it is a good idea to delay recalling him until the other, probably unfamiliar, dogs have left or been called off. In some situations, a dog perceives he is not allowed to move, or feels very unsafe doing so. In this case, it is not disobedience when your dog doesn't come when called, but self-protection.

When we teach, we often hear dog guardians complain about this type of behaviour. Typically, the dog simply lies down when another dog comes close and, no matter what the guardian does, he won't get up. Sometimes, it's the other way round, and the unfamiliar dog goes into a down as their dog approaches. When both dogs are fairly equal, it can happen that the one lying down suddenly jumps up and becomes aggressive. This behaviour is based on insecurity, not aggression, and stems from the dog's desire for more space.

When there is not enough space between the dogs, and flight is not possible (i.e., the dog is restrained by a leash), some dogs will switch from fear to aggression. Two major factors in this decision are the distance between the dogs, and the dog's previous experiences. In contrast, some lying down reflects the dog's decision to 'take what comes'.

It is important not to mistake lying down in this situation with a playful lying down – for example, waiting for a friend to get close enough to chase, or lying down while stalking pray.

SOLUTIONS

- You could ask your dog to sit down, with his side towards the dog that is lying down, and turn his head away. This breaks the front-on approach and is often enough to encourage the other dog to get up and continue on his way.
- You can walk around the dog who is lying down in a large arc to create more distance. This helps your dog as it prevents him from getting involved in a potentially impolite, face-to-face confrontation, which he may not know how to avoid.
- If there is not enough space to walk in an arc, you can turn around and find another area that allows you to create the distance the dogs need.
- If your dog has a tendency to lie down, it is best to turn away before he goes down to avoid the situation. In your dog's mind you have now taken care of his safety, which creates a stronger level of trust.

WHILE OTHER DOGS ARE APPROACHING

Many dogs will first lie down and wait to see what happens when other dogs are approaching them.

This female Labrador stops in her tracks as a group of seven dogs is approaching.

She then lies down and also rests her head on the ground.

When the dogs get closer, she raises her head and upper body. She remains motionless while Ella investigates her face thoroughly.

Even after the dogs left, it took quite a while for the Labrador to move again. She wasn't frozen in fear, but it seems she decided it was safer to stay lying down so she wouldn't risk an attack from such a large group of dogs.

As encouragement

Confident dogs will lie down to encourage an insecure dog to come closer. They will often avoid eye contact which would stop the nervous dog from approaching.

Here the two male Shelties, Jack and Calypso, are displaying a very polite and friendly greeting towards the young and insecure dog who is coming towards them.

This is the first time that these dogs meet each other. While Jack and Calypso move slowly, the other dog decides it is safer to flee at top speed.

As the Shelties and their guardian continue on their walk, the other dog decides to to come towards them again. As Jack and Calypso approach her, she stops immediately. In response, Calypso sits down.

Calypso turns his head away from her and Jack lies down. The other dog walks past the two Shelties in a large arc. Jack and Calypso have used the signals of sitting and lying down to tell the other dog that it is safe for her to come through.

To signal peaceful intentions

Shokran, the female Golden Retriever, is being greeted by a group of dogs. She has met the dogs before, but that was quite a few months ago, and she was very young then. We cannot assume the dogs recognised each other right away.

Shokran lies down and cowers. She keeps her head down and mouth closed. She looks at Arwen, who is approaching from the left, from the corner of her eye (*see inset*). Her movement is slow and minimal.

Shokran briefly got up, but when Tamika and Ella sniffed her intensively, she immediately lay down. Shokran often uses lying down as a communication signal when she is not feeling confident.

This scene played out just before Shokran opted to lie down. She is first greeted by Arwen and then her mother, Rani, and Tamika. When the meeting turns into an investigation, Shokran tries to de-escalate the situation by sitting down. She drops her head, lifts her front left paw, and pulls back her ears.

Shokran's topline is rounded and her centre of balance has clearly shifted towards the back. Her elbow is angled and her tail is flat on the grass, lying away from the body in a straight line. This tells the other dogs that she is communicating, and she is not afraid.

TO SETTLE OTHER DOGS

These dogs met in an off-leash situation, and they don't know each other.

The German Shepherd, who is lying on the ground, is facing several dogs whose body language seems confrontaional. He is using the signal of lying down to calm the other dogs.

As the male Boxer examines the Shepherd, he makes himself smaller. His tail is clamped between his hind legs, he shows a whale eye, his ears are pulled back and his mouth is closed.

The young Shepherd is getting a break because the Boxer is distracted by something else. He stays lying down, but now moves his head and opens his mouth.

TO SHOW RESPECT

Dogs have a lot of different ways to show their respect. Lying down is one option, but it is not used by all dogs. Dogs who lie down, often use sitting as an additional means of communication.

The two females Luna (*left*) and Foxi (*right*) live on a horse farm in Canada where they enjoy a lot of freedom. Luna gets along well with all the animals on the farm and when Foxi arrived, she was very respectful towards her.

Foxi's mouth is closed and she carries her tail high above her back. Her ears are turned slightly to the side and her front legs look stiff and braced. It is hard to see from the photo whether she is trying to make herself appear taller. This is a subtle demonstration to show she is mentally superior to Luna, which is understood by the other dog. Foxi is very self-assured, as opposed to Luna, whose tail is low and, in addition, her ears are pulled back and flattened against her head. Luna blinks, shifts her balance back slightly and there is tension in her back.

Luna opts to lie down; she was not physically coerced by Foxi. Direct eye contact, combined with the supporting posture, can encourage this respectful behaviour. Lying down is a way of confrontation. This subtle signalling may be hard for us to pick up and, if seen, it is frequently misunderstood.

WHEN INSECURE, AND CURIOUS

Cautious puppies often show this behaviour when they see unfamiliar objects. They will usually move further away to create more distance. This will change when the dog gains more confidence.

The Golden Retriever puppy, Tristan, respects these chickens and ducks, but he is still curious. If you look at the first photo on its own, you might think he is stalking the chicken. In fact, he lies down as the chicken gets closer to him and sits up only when the chicken moves away from him. He also turns his head.

FOR JOY

In this situation lying down is not connected to calming the other dog, nor feeling insecure (*see Joy, page 612*).

The little dog, Sally, is always happy when she sees other dogs. She greets them enthusiastically by jumping up and licking them, and then lies down. Here she is meeting the very gentle Irish Wolfhound, Olli.

WHEN INSECURE OR TO APPEASE

When Rani was young she was very respectful of Molly, the cat. Here Molly heads straight towards her with a highly raised tail, accompanied by loud meowing.

When Rani starts to lie down, Molly stops.

Rani lays down to appease the cat, who is following her, and is obviously in a bad mood.

Her strategy works. Molly no longer follows her. She sits down and breaks eye contact by turning her head.

10. LIFTING A PAW

REASONS FOR LIFTING A PAW

The signal of lifting a front paw, either while standing or sitting, can be seen in a variety of situations. It is used on an approach, to signal the need for a break when playing, to calm, as a request for space, or to confirm good intentions when meeting someone new.

Lifting a paw should not be confused with pointing, where the dog also stands and lifts a front leg. Pointing is a trait seen in some gundog breeds, but you can also see it in dogs of other breeds when they have detected wildlife.

There is also a type of paw lift that expresses caution or curiosity. It indicates the dog's is alert and focused, and can be categorised somewhere between a paw lift and pointing.

As with all the other communication signals, it is important to observe all the other signals the dog is sending and include these in your interpretation.

PAW LIFT IN PUPPIES

You can see very young puppies lift a paw; it could stem from a movement used when they were nursing.

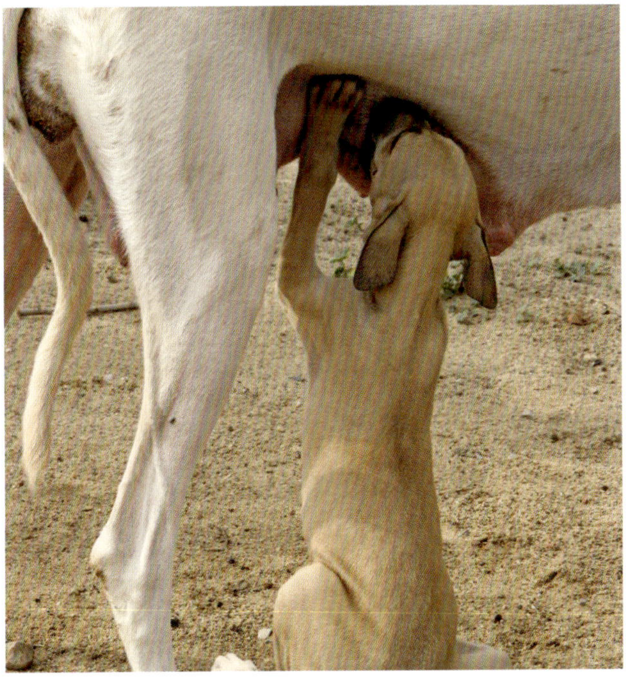

In this photo, you can see the Sloughi puppy trying to encourage milk flow. The milk ducts are stimulated by rhythmic kneading with the puppy's feet.

Here, puppy Alani lifts a paw to communicate. Tristan checks her out thoroughly. Alani is interested in her older half-brother and seeks a connection with him.

When puppies are insecure

Lifting a paw is used as a request for space. It can also express indecision when being approached.

This Sloughi puppy shows avoidance when a human hand comes close to him. He can't get away any further because he is already touching the wall. His face is very tense; his ears flat against his head and pulled back. He is lifting his paw asking for more space, or anticipating the discomfort he fears he will experience when he is too close to a person.

To calm a puppy and a mature dog

Here both dogs lift their paws with different motives.

Sloughi puppy, Chemsi, and Dobermann, Inca, meet for the first time. Inca is very excited; she is foaming at the mouth. In this photo, both dogs are very close and are looking directly at each other. Inca lifts her paw because she is insecure; Chemsi lifts her paw to calm Inca.

Inca lifts her paw every time Chemsi makes direct eye contact.

In addition to lifting her paw, Chemsi also blinks to calm Inca.

TO CALM ON AN APPROACH

Dogs that are over-eager when approaching other dogs can be calmed with a paw lift when they meet head on.

Arwen approaches the Yorkshire Terrier, who clearly shifts his weight as far back as he can. Arwen's hind end is still moving forward, but her front end is stopping in response to the Yorkshire Terrier's lifted paw, and to show she has friendly intentions towards the much smaller dog.

When both dogs are tense

When both dogs are very tense during a meeting, lifting a front paw can help to calm the situation.

We can see this is a tense meeting as both dogs are carrying their tails very high, Quivive (*left*) has her point of balance towards the back, her back is arched and her ears are pulled back. She is about to turn away from the male Poodle because there is not enough room for her. The male Poodle would like more connection; he is facing her and lifting a paw.

As encouragement

It doesn't matter if the dogs are small or large, lifting a paw encourages the other dog to come closer if he is feeling insecure.

Meeting at the off-leash park: The female, tan Dachshund-cross (*left*) greets the young black-and-tan, male Dachshund very politely while maintaining some distance. She is very interested; her tail shows her confidence and her ears are pointed to the front in an attentive manner. She lifts her paw and doesn't move and, therefore, does not invade the smaller dog's space. When he finds some courage and also looks for contact, she turns her head.

The female Dachshund cross does not move, which you can see in the positioning of her hind legs. She keeps looking away as the small male Dachshund comes closer and dares to sniff her hind end. Now she turns back to him. She lifts her front paw and turns her head to encourage the insecure young male to respond to her request for contact.

TO REQUEST A BREAK

While playing

Lifting a paw can slow down or end a game when both parties understand each other.

Quivive (*left*) and the Jack Russell-Beagle cross, Coco, meet for a wild game in the snow.

When it becomes a bit too much for Coco, she asks for a break by lifting a paw.

AS A REQUEST FOR SPACE

Pushy behaviour can also be interrupted with a lift of the paw.

Helge (*left*) approaches the young female Pug much too fast, which the Pug does not appreciate. She lifts a paw to ask for more distance.

TO SHOW RESPECT

Young dogs frequently use paw lifts.

Puppy Rani approaches the mature male, York, with great respect. Her posture and lift of her right paw show that she has friendly intentions.

WHEN TOO CLOSE

On or off leash

Lifting a paw is often used when other dogs are too close. This can be when they are on or off leash.

Off-leash: The black-and-white male (*left*) is clearly uncomfortable with having his hind end sniffed. He lifts his paw, his back is arched, and his hind legs are bent as if he is trying to tuck his hind end under.

On leash: The white dog is concerned by the tan dog, who is heading directly towards him. He responds by lifting a paw. His head is slightly lowered, his tail is level with his back and his body appears tense.

NOT LIFTING A PAW, BUT POINTING

Hunting breeds, such as Pointers, indicate that they have seen or caught the scent of wildlife by raising a front paw (pointing). This occurs in other breeds as well.

Katja wanted to take photos of her dogs. Taschka was posing for a photo when saw a rabbit in the field ahead of her. She was undecided as whether to stay still for the photo, or chase the rabbit. Here she shows the typical posture for pointing – her centre of balance is forward and her ears raised high, and inwards. Her tail is very low and is moving.

11. SNIFFING

REASONS FOR SNIFFING

Sniffing is used frequently as a communication signal, and for many different purposes. When meeting other dogs it can signal neutrality, disinterest, friendly intent or politeness. Sniffing can be used to appease, or to calm, when a dog is feeling insecure.

Sniffing can be employed with such focus and intensity that it is difficult to tell if the dog is using it as a communication signal, or if he is sniffing to gather scent and information.

TO GATHER SCENT

Naturally, sniffing is also used to simply gather scent. In this instance, it is not connected with a communication signal.

Here Bruno is in search of some scattered treats.

We cannot be certain if these dogs are sniffing to gather scent, or to release tension.

Dogs like to use sniffing as a communication tool within a group.

TO STAY NEUTRAL

Sniffing is often used by dogs to underline their neutral posture.

There is a certain amount of tension between the dogs in the background, probably because their wrestling has become too rough. Rani, in the foreground, demonstrates a very neutral posture by sniffing the ground intensely. She does not want to become involved in the dispute between the dogs behind her.

Sloughi puppy, Chemsi, tries a variety of strategies to calm the excitable Dobermann, Inca. Here she sniffs intently at a tree stump to signal her neutrality.

AS COMMUNICATION

Sniffing is one of the clearest communication signals and, usually, the other dogs will respond in the same way.

Rani and a Sheltie (*coming from the left*) meet a Golden Retriever, who is a stranger to them. The Retriever avoids eye contact by lowering his head. As the dogs come close, he blinks and turns his head away. While Rani keeps coming closer, with a raised tail, the other dog starts to sniff. The Sheltie has turned away. Another Retriever is entering the scene from the left.

The new arrival (Retriever) starts sniffing. The stranger (*back of photo*) has stopped near the tree during this sequence of photos. Both he, and Rani, carry their tails at the same height.

TO REQUEST A BREAK

Just like lifting a paw, sniffing can be a request for a break.

These puppies were enjoying a good wrestling match until the Bulldog started to sniff in order to request a break. The other puppy understands the suggestion and answers by sniffing as well.

A POLITE REQUEST

Sometimes sniffing is used as a request.

Puppy Arwen makes herself comfortable beside the Golden Retriever, York, whom she adores. Tamika sniffs the ground around them. She would like to play with Arwen and is now asking if the puppy would like to join her. Arwen looks to the ground and it seems that Tamika has not noticed her move. However, Arwen is sending a message that she is not really interested in playing right now.

Tamika turns towards Arwen who is lying down and asks again, just to be sure ....

However, Arwen does not want to relinquish her cuddly spot beside York, so Tamika walks off, still sniffing.

TO DEMONSTRATE NEUTRALITY

In these photos, we not only see neutrality, but also an approach with the intent not to disturb the dog that is lying down. These two dogs live together in an outdoor shelter.

The female, who is standing in this picture, is facing the male who is lying down. She is sniffing the ground and getting quite close to him.

As he moves his head towards her, she continues to sniff, but changes the direction of her eye and turns her body away from him.

As a response, the male looks away from her. The female moves slowly away, sniffing the whole time. In this case, sniffing is used as a respectful and polite gesture towards the male. It is clear that sniffing is being used as a communication signal when you look at her whale eye.

TO APPEASE AND CALM

When two dogs meet and one or both sniff, it is all about communication.

Yorkshire Terrier, Eve, is a peace-loving dog who always chooses to avoid confrontation. She moves towards the young dog, Robbie, sniffing the whole time. She moves very slowly towards Robbie who can be impulsive at times.

Robbie keeps a close eye on Eve as she passes.

After Eve has passed him, Robbie gets up and follows her.

Robbie has learnt from Eve and he is now sniffing alongside her, which surely helps to calm him.

WHEN APPROACHING AN UNFAMILIAR OBJECT

When dogs show sniffing behaviour very intensively, it can help to TTouch the dog and place a trail of treats leading towards the scary object, or to go up to the object to stroke it, and talk to it, without moving it.

When something suddenly changes in the environment, making the dog feel insecure, he will often resort to sniffing.

A chair has been dumped in the forest, and Rani has not seen it before. She is curious, but cautious. Her strategy is to approach carefully while sniffing.

When the chair does not react to her approach, and no longer seems scary, Rani gets close enough to touch it with her nose in order to investigate.

A CAUTIOUS CAPPROACH

Towards a stranger (human)

This wild Tunisian puppy is very shy towards people, but he is still curious and tries to approach. Here he uses sniffing, not to explore the surroundings, but to help him to be neutral and approach at a certain distance. Even when he looks up after having sniffed intently, his attention stays on what's in front of him, in this case Katja. You can see that from his whale eye.

TO COPY

Sniffing can also be used to communicate a casual, unnoticed approach.

Puppies like to copy what they see, even though the Potbellied pig, Hotte, is probably grazing.

... and a young Quivive responds to the cat, Nina, by sniffing.

Puppy Rani sniffs with Heller the cat ...

A DAILY MEET AND GREET

Sometimes it is quite difficult to differentiate sniffing as a form of communication from sniffing to 'read the newspaper'. Often it is used as a combination of both, as you will see in the following photo sequence.

The large spotted dog on the left is very absorbed in sniffing a special spot in the sand. He is not letting the approaching male Golden Retriever distract him.

At first the Retriever walks past him, but then he turns back and approaches from behind.

The spotted dog marks the area and turns away.

As he walks past, the Retriever intently sniffs the marked spot.

The large spotted dog changes his mind, turns, and returns to the Retriever.

Then he turns away from the Golden Retriever who is still sniffing and walks past him. He is not that interested any more; after all, he has already thoroughly investigated the spot.

The Retriever does not change his posture when the large spotted dog approaches, but he keeps an eye on him while he is sniffing (*see circle*).

12. URINATING

REASONS FOR URINATING

Urinating, as a communication signal, is often confused with marking – still seen as a 'dominant' or territorial marking behaviour, especially in male dogs. However, marking is something dogs do to ensure their presence is noted; it is often an expression of wellbeing or exceptional self esteem. Urinating can also be used to release tension, either for the dog who is urinating, or to settle the dog in his presence. As usual, we need to look at urinating in context with the other body language signals the dog is sending, and the specific situation.

We have mostly chosen photos of females to illustrate this topic, but males urinate to release tension in just the same way.

IN ADVANCED AGE

Older males, like Ronnie, often cannot stabilise their body on three legs any more, so they prefer to urinate while squatting slightly.

Very young males will also squat to urinate, a posture they copy from their mother. Depending on their confidence and body awareness – and their size – male dogs will start lifting a hind leg to urinate as they mature.

TO RELEASE TENSION

This small, tan female feels insecure when she is a group of free roaming dogs. Her emotions flip between mild curiosity with a desire to make contact, and near panic.

Here the tan female releases tension by urinating. As another dog approaches her from the front, she blinks. The light-coloured male comes closer while she is still squatting.

Since she cannot make a quick retreat, the tan female sniffs her own puddle of urine without making eye contact with the male. The raised hair on her back shows her degree of arousal.

The male now lifts his leg and urinates without facing her. We don't know if he is marking or trying to tell the female that he has no intentions of coming near her.

A little while later, her curiosity wins and she is brave enough to make direct contact with the light-coloured male.

TO RELAX

At a first meeting

Maxi is a very happy female Weimaraner, but she is always a little insecure when meeting dogs for the first time. Even though these two dogs know each other, Maxi is showing some concern.

The raised hairs on Maxi's back, especially near her shoulders and hind end, are easy to see. Quivive licks and shifts her point of balance away from Maxi. She avoids Maxi's direct eye contact by turning her head slightly.

Maxi's ears are raised at the base, which causes wrinkles on top of her head. However, her ears are not turned towards the front, but slightly pulled back instead. She carries her tail level with her back.

After making brief contact with Quivive, Maxi goes to the side of the trail and squats to urinate. Afterwards she makes contact with Quivive again, and appears to be more relaxed. Her ears are now pointing forward, they are raised less and the top of her head is flatter. The hairs on her back are less raised.

TO MARK TERRITORY

Conscious placing of urine is often connected to marking.

The female Berger, Verveine, is visiting Gabi and Quivive for the first time. After a long ride in the car, she confidently walks into the garden. Quivive stands still and stiff, with her tail raised high and unmoving, while she watches Verveine.

After Verveine relieved herself following her long trip, Quivive searches for the spot and quickly marks very close, nearly on top of it.

Quivive has a habit of marking on walks, mostly in unfamiliar areas, or on trails she had not been on for a while. It is likely that her purpose is to document her presence.

AWKWARDNESS/EMBARRASSMENT

Much like scratching and yawning, urinating can be seen as a gesture of awkwardness when the dog does not really know what to do. It is also used as a means of splitting two dogs (*see Splitting, page 350*).

These two females meet and are now standing close to each other with direct eye contact. We observed that the dog on the left is older, and a number of other dogs had paid respect to her during the course of the day. She did not seem well physically; she moved with a stiff back and drooled excessively. In this photo it seems she has signalled to the female, standing to her right, that she needs more space. Her point of balance is shifted towards the other dog.

The other female retreats and squats to urinate The dogs are not looking at each other any more.

Now the younger female starts to scratch her neck. In this case we see the urinating and the scratching as a sign of awkwardness or embarrassment. The older female has shifted her point of balance back to the centre and remains motionless in the same position.

COMBINED WITH GROUND SCRATCHING

By a male

Scratching, or kicking dirt, after urinating is often observed in male dogs.

Quivive's sire, Laguiole, is a very confident male and he marks a clump of grass. He examined the spot carefully before choosing it, and is putting a lot of effort into making sure he completely covers the entire area.

He makes sure his actions don't go unnoticed by actively scratching the ground with his hind and front feet.

By a female

Ground scratching can also be observed in female dogs. But you rarely see them putting quite this much effort into it as this female is showing us.

TO FLIRT

Why dogs scratch the ground is not completely understood. Since wolves share this behaviour, it could be a genetically predisposed behaviour. Another possibility is that it spreads scent, as dogs have scent glands on their paws.

This Standard Schnauzer scratches the ground after he urinates – not just with his hind feet, but also his front feet.

He wants to flirt, which you can see by the way he holds his ears. However, his excessive facial hair which also covers his eyes, really irritates both Rani and Tamika. He turns his head and leaves.

TO APPEASE

Katja's dog family was just heading out for a walk when this photo was taken and, therefore, all the dogs are excited. The excitement at the beginning of the walk usually triggers the black dog, Taschka, to settle everyone.

You can just see that Arwen lies down in front of the black dog, Taschka, and Tamika is intent on sniffing the ground. Tamika's nose is just above the ground, and her eyes are almost closed, because she is facing Taschka front on. Rani is in the background and chooses a different strategy to relax the situation; she squats and urinates. Here we can really see the signalling within the group. Jazzman, the small Yorkshire Terrier in the foreground, sniffs a blade of grass. All four dogs around Taschka have their own individual communication signals to appease her and relax the situation. Taschka has turned her head away from Arwen and is looking at Rani.

Did you know that females can also lift a hind leg while urinating?

This is Quivive urinating with one hind leg lifted, a habit that can be observed in numerous females.

13. SPLITTING

REASONS FOR SPLITTING

Splitting is the term used when a dog takes on the task of splitting two dogs who are too close, so as to create more distance between them. The dog who instigates the split is not involved in the original conflict or tension.

There are different reasons for this. It may be that one of the dogs needs to be protected; the game may have become too rough, or one of the dogs may be overwhelming the other. Depending on the 'splitting' dog's level of confidence, the dogs involved can be split very quietly, simply by walking between them. However, if tensions are running high, more vehement action may be called for.

Unfortunately people often misinterpret the 'splitting' dog as being jealous. Jealousy is definitely an emotion dogs will show (*see the photo sequence with Tachka, page 353*), but it does not apply to this situation.

TO PROTECT

While out for a walk, Eve, the Yorkshire Terrier, makes a connection with a Rough Collie she has never met before. Since Eve rarely seeks out contact with other dogs, Arwen feels responsible and wants to make sure everything goes well. She approaches with a slightly lowered head and ears back, a typical posture for her that shows a mixture of humility and caring.

Arwen creates space between the Rough Collie and Eve by pushing her way between them. You can see by her tense, and tightly wagging tail that she does not feel very confident.

Eve was fine during this interaction and it does not appear as though the Rough Collie was a threat to her. However, Arwen felt it was important to interrupt the interaction.

TO CREATE A BREAK

Some dogs feel the need to speak for others and use the splitting technique to interrupt a tense or stressful situation and create a break.

The black Labrador Retriever (*right*) wants to support the small female Cocker Spaniel, who needs a break following a wild game with the black mixed breed male. The Labrador keeps walking between the two recovering dogs.

He keeps protecting the small Cocker, who is showing the mixed breed male that she still needs a break from him.

A POWERFUL PROTECTOR

When splitting, most dogs don't differentiate between dogs they know and those they don't know.

The black dog, Taschka, approaches the young female Golden Retriever much too fast (*see arrow*). Spot turns immediately.

In this photo you can see why he turned so quickly. He wants to protect the young Retriever from Taschka's over-bold approach.

Spot pushes himself between Taschka and the Retriever, who is clearly worried. Spot looks at the Retriever with a kind, friendly expression. He leans away from her and blocks Taschka.

After stopping Taschka's advances, Spot stays close by to keep an eye on the situation.

Splitting is a strategy that we, as dog guardians, should employ to protect our own dogs, or other dogs that may be involved. If your dog feel safe in your presence, it will strengthen the trust and bond you share. Bear in mind, feeling safe is a top priority for all dogs.

However, if a dog moves between a person and another dog, it can become problematical, especially if the dog's guardian is involved. In this situation, it is best to enlist the help of a professional trainer or behaviourist.

ACTING ON IMPULSE

This beach is a popular off leash area; dogs who don't know each other frequently meet and interact.

In the following scenario Taschka, the black female to the right (*circled*) and the light-coloured Arwen, in the foreground, are members of the same dog family. The female Rhodesian Ridgeback's intention is not completely clear as she charges towards the dogs, but Arwen is very uncomfortable with the situation.

While Arwen is getting into trouble, it seems that Taschka (*the back female*) is more interested in the black Labrador who is in the water.

Arwen avoids the Ridgeback by moving into the water. The Ridgeback stops and shifts her balance far towards the back. When Taschka realises that Arwen is feeling threatened, she changes her direction and heads towards the Ridgeback.

Arwen carefully approaches the Ridgeback who shows that she is not interested in a confrontation by turning her head away.

Taschka is unstoppable; she wants to halt the interaction. She goes after the Ridgeback first, but then turns on her friend, Arwen.

Taschka is not happy until Arwen turns away from the stranger and leaves. Tashka feels she has an important role to play. Arwen is part of her family, and is not supposed to make connections with other dogs. Taschka's actions seems quite aggressive – they certainly made an impression on the Rhodesian Ridgeback – but there is no physical contact between the two dogs, which are fairly equal in size. Taschka's point of balance stays in the centre, and the Ridgeback is leaning away from her.

There are some dogs who are on a mission to interrupt interactions between other dogs. However, not all dogs appreciate, or accept , this type of intervention. Often it is the dog who is doing the splitting that gets into trouble. Ideally, guardians should recall dogs, using a positive tone of voice, to prevent the situation escalating.

Other dogs use the splitting strategy because they feel insecure; they want to stop every game because they lack confidence in canine interactions. Gabi's Fanny was that type of a dog and usually had to be leashed when she met other dogs on a walk, so the others could play uninterrupted.

SUBTLE INTERFERENCE

In this sequence of photos, a wild wrestling game between two young Golden Retrievers turns into a a skirmish. Some of the older dogs come closer to watch the action.

The youngster standing over the other young dog is very excited. The adult male (*with the blue checkered handkerchief*) approaches politely and sniffs the aggressive dog's rear end. He stays away from the angry youngster's teeth.

The youngster lets go of his 'victim' and makes direct eye contact with the dog who is standing in front of him. That dog immediately turns his head away. The young dog flicks his tongue, which could have been triggered by the older male who sniffed his rear.

Another adult male (*with the red handkerchief*) approaches the group. The first male with the blue checkered handkerchief continues his intense sniffing and the youngster, who was fighting just a moment ago, is letting go of the dog underneath him.

The mature dog (*with the checkered handkerchief*) pulls back a little and watches the scene. The dog (*with the red handkerchief*), who has just joined the group, is trying to make eye contact with the troublemaker, who is starting to wrestle again. The youngster blinks.

Now the older dog (*with the red handkerchief*) takes action because the youngster is not responding to him, and keeps attacking the dog underneath him.

The subtle interference of the two older dogs finally results in the excited youngster leaving his wrestling partner, which gives him a chance to get up. The two mature dogs were able to defuse the situation in a very calm and confident manner. Of course, it would have been even better if the boisterous youngster had been removed from the situation by his guardian as soon as the conflict erupted.

14. BOWING

REASONS FOR BOWING

This posture is universally used to invite play and is, therefore, seen as a happy, positive behaviour However, bowing is also used for communication purposes, triggered by an emotional response. It can be used to show insecurity, or to reassure a nervous dog that the intentions are peaceful without wanting to play. As we have mentioned with other communication signals, we need to evaluate the entire situation that triggers the dog to bow.

INVITING PLAY

When a dog invites another to play, he will often have fun facial expressions and movements will be exaggerated.

The female Rhodesian Ridgeback mix, Nelly, invites Maja to play with her. Maja responds by bowing as well.

An invitation to play, combined with a bark.

In profile

The ears are loose, held towards the back and slightly raised. There is direct eye contact and the weight is shifted far back with extremely bent joints.

Front-on

Typical posture to invite play: The ears are held in a funny position, to the side and flopping, the eye is soft and, in this case, without direct eye contact towards the other dog. The face is relaxed, the front legs spread far apart. The front end is quite low with strong angulation of the elbows. The hind end is noticeably higher and the tail usually wags and is held just slightly above the level of the back.

SHOWING INTEREST

While flirting, or when interested in another dog, many dogs will bow.

The small male Yorkshire Terrier, Champs, invites Lola, who is on leash, to play. His elbows almost touch the ground, his hind end is raised up high, and his point of balance is towards the back. His ears are turned to the side and back and they are raised; he is making direct eye contact. This is a typical flirting face. He likes Lola and his invitation to play is more an expression of his interest in her.

TO ENSURE PEACEFUL INTENTIONS

A confident dog will often make himself smaller to encourage the other dog to play, and to confirm that his intention is to play and to be kind.

Spot makes himself extremely small by lying on his side to invite the puppy, Robbie, to play with him. In this way he reassures Robbie that his intentions are friendly, and so the small dog is more likely to engage in play.

SIZE DISCREPANCY

Sometimes a bow indicates a conflict between curiosity and insecurity.

The small, dark-coloured male (*left*) and the large tan-and-white male (*right*) meet the Silken Windsprite female, Emmie. The small dog's bow shows he is undecided, his ears are pulled back and he is holding his tail very low. His point of balance is far towards the back, and a little to the side, as he tries to increase the distance between Emmie and himself. He maintains eye contact with Emmie. She approaches him with a low head carriage and her tail is level with her back. She may be a little concerned by the small dog's pronounced facial markings.

Now all three dogs take off – the small dog keeps a certain distance from Emmie by leaning away. Only his head and neck are turned towards the two larger dogs. Emmie and the large male are looking at each other while running.

INDECISION IN WILD DOGS

Katja met these puppies belonging to a family of wild dogs at the edge of a desert in Tunisia. They are very reserved, but still curious, which is the case with most young animals.

The play bows shown on these photos are not really an invitation to play but, more likely, an expression of ambivalence. Katja, who is standing still while taking photos, is approached by the curious pups. When they get too close to her, they start to bark or to take evasive action. Note the whale eye in the puppy who is jumping away, and keeping a close eye on Katja, who does not move from the spot.

In this sequence of photos, you can see the puppy's emotions flipping back and forth:

- Curiosity with a play bow position
- Insecurity and fearful evasion to the side
- Brave with a tendency to attack
- Followed by evasion, head turning and back to a play bow.

TO CALM

Here the young Eurasian, Diana, is using a bow as an invitation to play, but it is also used as a calming signal.

Play bow to calm

- Tail held to the side
- Front end low
- Front legs far apart

Diana tries very hard to encourage Quivive to play using play bows.

Quivive threatens the four-month-old puppy, Diana, with exposed teeth, because the youngster is too boisterous for her. In response, Diana uses a play bow to calm Quivive who clearly needs it. Diana wants to express her intentions for a peaceful play.

Diana keeps a safe distance from Quivive, with her point of balance far to the back. She makes herself small, and her ears are pulled back.

When Quivive first met Diana on a walk it looked as though she was willing to play. This was not the case, and, with hindsight, Quivive made the right decision. The youngster was physically stronger, and very boisterous; if play had got underway, Quivive would have been overwhelmed.

Once Quivive set the ground rules, the dogs kept their distance from each other for the duration of the walk, even though Diana made repeated requests to play. Since Quivive showed so clearly that she was not interested in playing, Diana had to be put on the leash a few times during the walk. During these on-leash spells, Diana's guardian distracted her with fun, positive activities. After repeating this strategy – putting her on the leash and then releasing her – Diana started to sniff with Quivive, run on her own, and remain neutral towards the older dog.

TAKING RESPONSIBILITY

Young dogs often lack the ability to assess a situation, and this applies most particularly to their interactions with older dogs. They fail to show respect and keep pestering with on-going requests to play. In most cases, the youngster does not have complete control over his body and, therefore, appears unbalanced. In these situations, the guardian should step in and calm the youngster's unruly behaviour.

Mature and older dogs know from experience how tiresome it can be to fend off a young dog who is allowed to act without jurisdiction. They have good reason not to accept every invitation to play. Their reluctance may be due to physical ailments, problems with their joints, a chronic illness, heart problems, insecurity about contact with dogs in general, or simply the desire to be left in peace. Much like people, old dogs simply don't have the same desire to play as young dogs.

There are very few older dogs that enjoy playing with unfamiliar dogs. If the dog retains a desire to play, he is more likely to engage with the dogs he lives with, or those he has known since puppyhood.

TO ENCOURAGE

Measured invitations to play, without exaggerated signalling, are used by dogs who are more reserved and require gentle encouragement.

Rani (*front*) meets a young, slightly nervous dog who is holding his ears to the back. The hair on his back is raised. Rani is clearly inviting him to play.

When Rani realises that the young dog is receptive to her invitation, she makes herself even smaller. She doesn't jump in his direction right away, as she would do with a more confident dog. Instead, she waits. The young dog lifts a paw and then play bows.

15. JUMPING AWAY

REASONS FOR JUMPING AWAY

This refers to a wild, crazy jump, which is usually seen in a very tense situation. When a dog jumps in this way, it is a means of decreasing his own tension, or as a signal for the other dog to disengage. It may be used when a dog is feeling indecisive, and it can also be an invitation to a game of chase.

TO RELEASE TENSION

In a tense situation

There are a lot of wild street dogs in Tunisia who mostly loiter at the edge of the villages where they can retreat into the desert if people come too close to them. These dogs live in large, sociable family groups.

Inca, a female Dobermann, who is a family pet, meets a family of wild dogs with young puppies, while out for a walk. The adult dogs clearly warn her not to come any closer.

Inca makes eye contact with the light-coloured female. Her ears are held to the side and raised at the base, her mouth is closed. The tan-and-white female seems to sense the beginnings of a conflict. She starts a very silly looking action; her aim is to decrease the tension.

Inca now flicks her tongue and shows a whale eye (*see inset*). The light-coloured female has her tail raised high. The tan-and white dog jumps away from Inca and the light-coloured female then turns away.

Here you can see the jump away in action.

When Inca turns away and continues on her way, the situation relaxes.

TO RELIEVE PERSONAL TENSION

Micky, a small mixed breed, meets the very casual Golden Retriever male, Daniel, for the first time.

Micky jumps towards Daniel who does not change his posture. When Daniel moves a little bit, Micky dashes off to the side.

Micky's aim is to create some distance and then run at the larger dog again. Just before he gets close, he turns away at great speed.

Daniel relieves the tension with a silly leap to the side. Micky evades sideways. In fact, Daniel has barely changed his position; he shifts his weight back on to his hind end as soon as he finishes his sideways jump. He wants to show Micky that he has no interest in a conflict, but he does not want to issue an invitation to play.

NEUTRALISING AN OVER-BOLD APPROACH

With an invitation to chase

Young dogs sometimes run towards other dogs without much control or consideration. Here the female Silken Windsprite, Emmie, jumps to the side to invite a game of chase.

When they meet at a distance on the beach, the Borzoi gives a very clear signal that he wants to play with Emmie. However she approaches at full speed and, as she gets closer, he rises up to his full height and breaks direct eye contact. Note the slight whale eye (*see inset*) which shows he is keeping Emmie in his field of vision.

The Borzoi does not change his posture, except for his head, which is not fully turned towards Emmie, allowing him more direct eye contact. He has shifted his weight towards the back. Emmie returns his eye contact, her ears are pointing forward and up and her front legs are far apart in a typical invitation to play. She responds to the Borzoi's invitation to play with a jump to the side and asks him for a game of chase.

SURVIVING A TENSE SITUATION

This meeting has caused one of the dogs to freeze. A wild jump to the side relieves the physical and mental tension.

The small black dog, Taschka, makes contact with a chocolate Labrador. Her posture expresses self-confidence. The chocolate Labrador keeps his ears tight against his head and pulls them back. He appears to be frozen. The black Labrador, standing on the right, feels the tension but wants to remain neutral. He looks in the other direction and then turns away from the two dogs. The chocolate Labrador remains motionless until Taschka leaves.

When Taschka is at a safe distance, the chocolate Labrador wildly jumps to the side. This helps him relieve the tension that has built up in his body.

FOLLOWING A PRESSURISED GAME OF CHASE

When one of the dogs feels that he needs more space while playing, a strategic jump to the side can give him the distance he needs

The Elo (*left*) and the small russet-coloured female are engaging in a game of chase. The female's tail is clamped and her back rounded. The Elo has caught up with her and there is no longer enough space between the two dogs.

The russet-coloured female resolves the situation by adding a crazy jump to the side, thereby giving herself more space. Her tail is almost touching her belly. The Elo comes to a standstill and lets her go.

WHEN STARTLED

Jumping to the side is also a common response when startled.

The Golden Retriever, Leonito, and female Weimaraner, Deluxe, are standing together on the shoreline. Deluxe is looking at Leonito. His head is turned towards her, but we can't see what he's looking at.

As Deluxe suddenly starts barking, Leonito leaps to the side and takes off. We don't know why Deluxe started barking, but Leonito quickly released the sudden tension with his big leap.

RELIEVING TENSION

Created by close proximity

Even if these dogs were friends, tension can still arise when there is no space between them.

The female Briard, Chocolat (*right*), makes contact with the male. Both dogs hold their ears in the flirting position; they are very close to each other.

Chocolat does not usually tolerate dogs that close to her, cannot handle the tension any longer and jumps to the side. It seems like an invitation to play, but you can see from the male's response that he understood the message correctly. He turns his head and does not follow her.

16. SHAKING OFF

REASONS FOR SHAKING OFF

Dogs usually shake, or shake off, after a difficult situation. It is as if the dog wants to shake the stress and tension from their body. We may also see the response when a dog is happy or excited.

Sometimes a dog stops mid-shake. If there are no direct external influences, such as a visual distraction, the dog could be experiencing pain, particularly if he does this regularly. Older dogs, who are not as stable on their feet, may show either an incomplete shake, or they may give up the behaviour entirely.

Dogs shake from front to back and, usually, the head has finished shaking when the hind end is still moving.

FOLLOWING HIGH AROUSAL

A confrontation is the cause of excitement and arousal. Shaking off relieves the physical tension once the interaction is at an end.

During a Retriever-hike, the group meets an unfamiliar male (*between the two Retrievers with the red handkerchiefs*). The three entire males escalate the situation, and the tension rises (*see page 401*).

The three males separate from the group and begin to settle down.

The stranger returns to his guardian and shakes off the tension following the high arousal encounter.

TO RELIEVE TENSION

The use of a life-size stuffed dog can be very useful when teaching meet and greeting manners, as most dogs see them as real dogs.

This male Rhodesian Ridgeback has problems approaching other dogs. Here he gets to meet a stuffed dog who was placed in a neutral position with his side towards the male. The Rhodesian Ridgeback is very tense as he approaches. After he has confirmed that the stuffed toy is 100 per cent neutral, he shakes to relieve the tension.

A dog will often shake when another dog has been too close to him. This happens just as often between dogs who know each other.

Quivive, who is a puppy, pursues Tiber, who is blind.

Quivive is too rough with Tiber who can no longer defend himself.

When Quivive is recalled by Gabi, Tiber realises he will now be left alone. He shakes to relieve himself of the built-up tension.

When these males meet in the forest, it seems that they are quite impressed by one another. Initially Bruno, the dark tan male, appears intimidated. His ears are pulled back, the hair on his back is raised, and his tail is carried low – all signs that show he is uncomfortable. However, he then faces the black Labrador, .

As soon as the other male leaves, Bruno shakes off his excitement.

The following photos were taken when Rani was still quite young. They record a meeting with the cat, Heller.

Rani has laid down very cautiously, but Heller air-licks. This was a cat who was very familiar with dogs and, therefore, used the same communication signals. When Heller decides to walk past, Rani gets up and shakes off the tension. Heller blinks as he walks past her.

Even when the situation has not been overly stressful, many dogs will shake afterwards.

Here a white German Shepherd passes a Golden Retriever who is sitting down. The Retriever's ears are held flat against his head, he has his head turned away slightly, and he blinks. The Shepherd turns his ears to the side and his eyes are small. He carries his tail a little higher than his back and it is moving. Both dogs seem a little tense.

As soon as the German Shepherd has passed, the Retriever gets up and shakes thoroughly.

AFTER MAKING CONTACT

Shaking as a communication tool is usually displayed by all dogs involved in a situation, one following another.

Here two dogs, who are not familiar with each other, make a careful connection.

They then shake to communicate, first one, then the other.

AN EXCITED MEETING

This behaviour is part of the gestures of belonging repertoire, *see page 456.*

Quivive shows a lot of different communication signals when she is excited to meet someone she loves; shaking off is one of them.

AFTER A COLLISION

Here we witnessed an accidental collision between these two black dogs after one of them had been jumping back and forth excitedly. The casualty of the collision did not care; he was completely focused on the dog behind the fence. The black female in the foreground, who caused the collision, shook immediately afterwards to relieve the tension.

A CONTROLLED MEETING

Following a short meeting, dogs will often shake.

In this sequence of photos, we can see that the female Rhodesian Ridgeback is tense as she is uncomfortable with the Weimaraner sniffing her rear end. She shakes while the Weimaraner is sniffing her. In this way, she may be endorsing her option to stay neutral. After the shake off, she walks away with her tail held high.

AN INDECISIVE APPROACH

Shaking can also release tension when meeting another dog during a walk.

During a walk, this Bull Terrier approaches Gabi and Quivive. His ears are raised high and pointing forward. His forward movement seems very tense. Quivive moves behind her guardian and sniffs the side of the trail. As a response, the Bull Terrier starts to shake while still approaching. His aim was to relieve the tension he was feeling.

17. SCRATCHING

REASONS FOR SCRATCHING

Scratching is generally a gesture of embarrassment or awkwardness. However, dogs can also be affected by a variety of skin ailments or parasites, so it is important to evaluate the overall situation. It is worth bearing in mind that emotionally charged situations can also increase the sensation of itchiness if the dog has parasites.

Puppies will often scratch at their collar or harness when they are not used to wearing it.

WHEN INDECISIVE

Someone is approaching this puppy and he can't decide how to respond. He would like to make contact, but he is very shy. As a puppy, one might assume he has fleas, but this is not the case.

AN UNFAMILIAR ENCOUNTER

Puppy, Arwen, meets the Potbellied pig, Lucia, for the first time and she feels quite insecure.

She leans away, but cannot look away. She has to keep an eye on the strange creature.

As Lucia turns towards the puppy, Arwen sits down. She keeps on scratching until Lucia turns away.

WHILE PLAYING

Beagle mix puppy, Todt, is playing with Honey, the cat.

All of the sudden, Todt is not so sure he wants to continue the game – maybe he had a taste of Honey's sharp claws.

Todt interrupts the game and starts to scratch himself. He turns away from Honey while he scratches, and doesn't stop until the cat has also turned away.

18. YAWNING

REASONS FOR YAWNING

Yawning is another of those behaviours that we label as a sign of embarrassment, but this is not strictly accurate. In our experience, yawning is also used, and understood, as a communication signal. This may be on a sub-conscious level in some cases.

Yawning is often shown when dogs are emotionally indecisive, when they want to calm themselves or another dog, in a stressful situation or when they are in pain.

Of course, dogs also give a hearty yawn when they are tired, or are just waking up.

Movement of the tongue while yawning.

FEELING INDECISIVE

This dog lives in a shelter where the dogs are kept in large outdoor spaces in groups. As a former street dog, he probably did not have a lot of positive interactions with people in the past, but he is wanting to be close to Gabi.

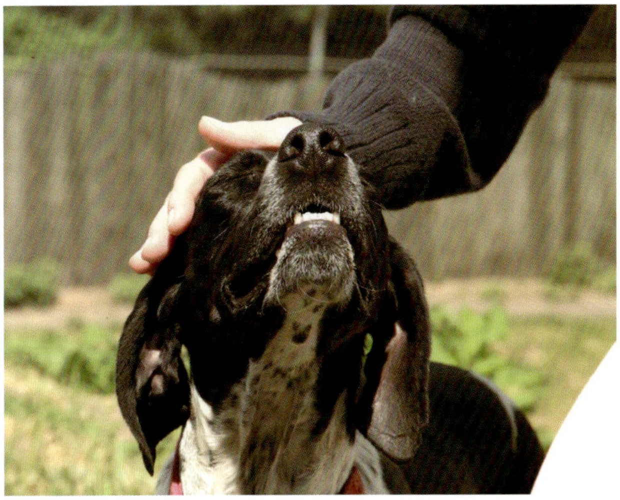

This type of contact makes the dog feel uneasy, which he shows through his yawn. However, he stays in place and allows the touch because he is also enjoying the attention.

DUE TO RESTRICTION

Caused by a poorly fitted harness

It is very important to make sure a dog's harness fits correctly, especially in the case of a fast-growing puppy. The discomfort caused by a poorly fitted harness, or a collar that is too tight, can result in an increase in yawning, licking and air-licking.

This harness has slipped below the dog's left shoulder blade, which clearly affects his ability to move.

WHILE RESTRAINED

A group of dogs is approaching, but this dog is unable to move because his guardian is putting on his harness. He signals his sense of conflict by yawning. He remains still, even though the procedure is uncomfortable. It is likely that yawning has a calming effect on him, in much the same way as yawning helps us to relax.

Many dogs, including those that did not grow up on the street, dislike having their harness put on. This can be due to an uncomfortable fit, or if the harness has to be put on over the head or requires the front feet to be picked up individually and placed into the harness. This can be very difficult for dogs who have poor balance or are in pain.

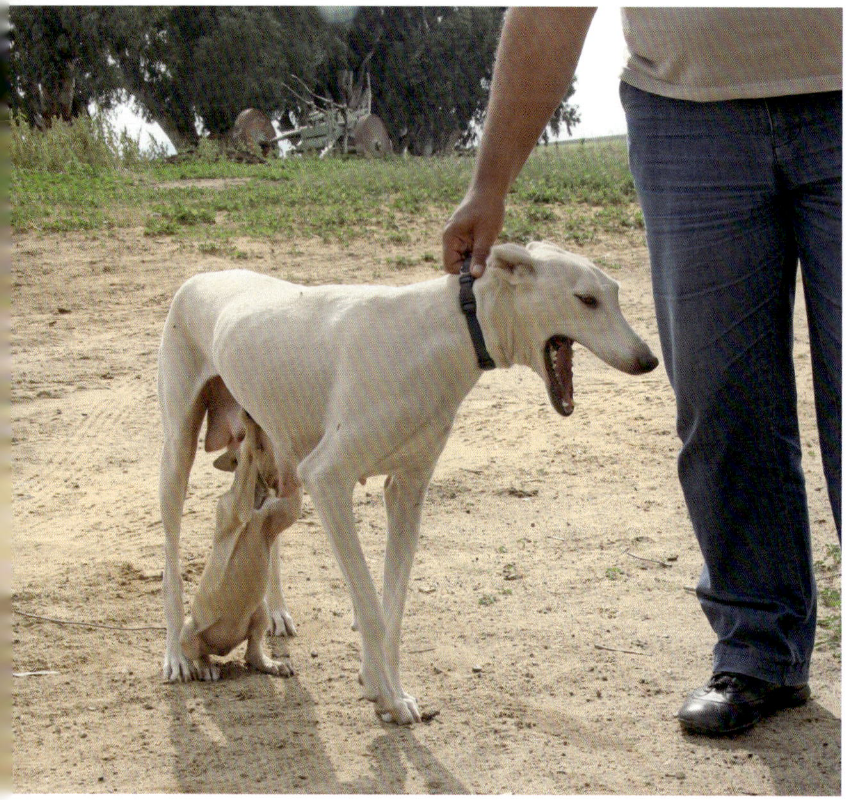

This female Sloughi lives in a breeding facility in Tunisia where she has very little contact with people. Here she is being held by a care giver so she can nurse her puppies. She is uncomfortable with the hand on her collar, either because it inhibits her ability to move away or because the person is too close to her. It is also possible that she is yawning because a stranger is pointing a camera at her, or simply because the sharp puppy teeth are hurting her nipples.

We have observed that companion dogs, who are very used to contact with people, often don't like to be held by their collar or harness. Most dogs dislike this type of restraint and will either try to evade it, or try to communicate with signals such as tongue-flicking, crouching, blinking or yawning

If you see these behaviours in your own dog, it is helpful to practise these situations in a positive manner, or to relax them with the Tellington TTouch® method.

LOOK AT THE CONTEXT

Of course, dogs also yawn when they are tired, but yawning, shaking off and scratching are all signs of embarrassment gestures. Once again, we need to look at the context to evaluate why a dog may be yawning. We have observed yawning when the dog is emotionally indecisive, for example while meeting other dogs, or when he doesn't understand what is expected of him. Yawning is often triggered in a training situation when a handler is giving cues, or behaving in a manner that is not understood by the dog.

Dogs use this signal on a very individual basis; some use it frequently and others rarely. It may also be used in combination with other signals, such as turning the head away.

Finally, yawning may be an indication that a dog is in pain. We have had a number of veterinarians confirm this observation, so it is important to mindful of this when assessing your dog's behaviour.

TO RELIEVE TENSION

When a dog yawns it not only relaxes the facial area but, to some extent, the whole body.

This female lives in a group in a shelter. She starts to yawn as some of the other dogs in the group approach her. After the yawn, her face is more relaxed, her ears are not held back as tightly. To better understand the emotions of this dog, *see page 91*.

TO SELF-CALM

Yawning is not only used to calm others, it is also used to self-calm.

The adult Border Collie, Luke (*right*), has a lot of issues with the youngster, Jake, who has recently joined the family. Here both dogs are lying down, facing each other. Both are able to look away.

As Jake turns his head towards Luke, the older dog starts to yawn. In this case, he is yawning to calm himself.

A RESPONSE TO PAIN

Yawning is often shown when a dog is experiencing pain. This can be the case when a dog is physically incapacitated due to age or injury, but it can also be triggered if you make contact with a painful area.

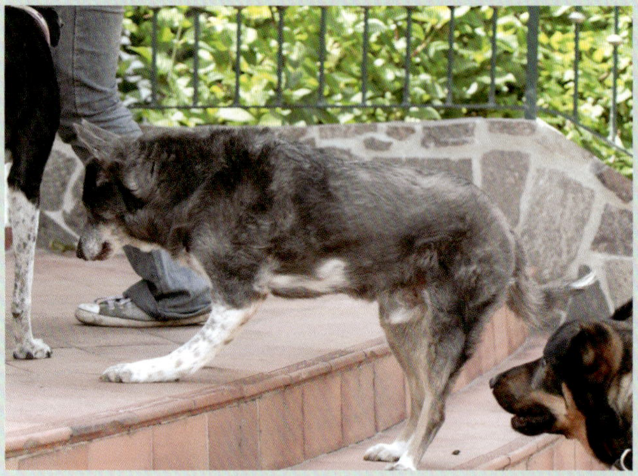

This old dog lives in a shelter. He is blind and his hind end is extremely stiff, so much so, he can barely bend his joints. His front legs are crooked, possibly the result of poorly healed fractures. Walking up the stairs is very difficult for him and obviously very painful. He yawns as he tries to negotiate the stairs.

PART THREE:

THE EMOTIONS

OUR DOG'S EMOTIONS

Dr. med. vet. Daniela Zurr, Tellington-Instructor for dogs and small animals, specialising in veterinary behavioural therapy

As human beings, we are often very quick to interpret our observations of other people and animals. While doing so, we automatically assume that the other being experiences similar awareness and emotions to our own. However, we only need to consider the dog's sense of smell to recognise how differently they perceive our shared environment.

Emotions are expressed through body language and mimicry. These have been developed through evolution and apply to all species. Humans, as opposed to other species, attach names, labels and categories to the emotions, such as fear, anger, spitefulness, joy, sadness, and disgust. These are dependent on the individual's experiences and the cultural circle in which the person was raised. The names, or labels, hold a slightly different meaning for each individual. We, therefore, need to remember that when we put ourselves into another person's place, it can only be an approximation.

If this is applicable to people, we have to be even more careful when we try to interpret the emotions of other mammals. We need to start by setting an objective baseline for our observations. By studying the photos in this book, thus far, we have had the opportunity to observe the dog's expressions and body language and to be mindful of the context in which those signals are displayed.

To determine how to interpret these observations, and to form a conclusion, we need to go one step further. We need to become skilled at predicting the dog's next move, which could be within a few seconds or a matter of minutes. The authors of this book have given us the tools. However, to be successful, we must resist the temptation of putting dogs and human emotions in the same category. This may be useful for human-to-human communications, but it can only ever be an approximation of what a dog is experiencing.

1. RANGE OF EXPRESSIONS

To discover the range of emotions that dogs experience, we are focusing on one dog, Quivive, the Pyrenean Sheepdog. In the following series of photos, notice the position of her ears, the shape of her eyes, the position of her eyebrows, her mouth, and especially the corners of her mouth and tongue as she expresses different emotions.

RELAXED AND CONTENT

While going for a walk

- Eyes almond shaped
- Eyebrows neutral
- Ears slightly raised at the base
- Tips of the ears falling down loosely
- Mouth slightly open
- Tongue loose inside the mouth

ATTENTIVE AND CONTENT

Waiting

- Eyes slightly rounded
- Ears pointing forward
- Tips of the ears falling forward
- Mouth closed
- No tension in the upper lip
- Hair around the mouth and nose flat

SAD – DEPRESSED

Experiencing a false pregnancy

This photo was taken when Quivive was feeling extremely depressed, experiencing a false pregnancy. It was very difficult to motivate her to do anything.

- Eyes rounded
- Eyelids hanging
- Eyebrows pulled down
- Ears hanging back
- Hair of the head and face hanging
- Mouth closed

FEARFUL

The first time Quivive met a Potbellied pig

- Eye very round
- Ears held tightly to the back
- Hair on top of her head is held close – the top of the head is tensed
- Mouth closed
- Upper lip tense

VERY RELAXED

Following a Tellington TTouch® session

- Eyes noticeably almond shaped
- Ears held back
- Tips of the ears hanging down loosely
- Hair on top of the head fluffy – no tension on the top of the head
- Mouth closed
- Hair is loose around the nose and mouth

HAPPY AND EXCITED

Extreme excitement when Gabi's daughter, Tina, comes to visit

- Eye slightly rounded
- Ears held flat and back
- Mouth open
- Corners of the lips pulled back – long mouth
- Lower lip hangs loosely – teeth visible
- Tongue muscle raised
- Tongue inside the mouth

LOVING AFFECTION

Connecting with a much-loved guardian

- Eyes rounded
- Ears pointing forward at the base
- Tips of the ears hang down loosely
- Mouth slightly open
- Upper lip slightly raised
- Lower canine incisors visible
- Corners of the mouth pulled upwards

LOVING AFFECTION

A cuddle with a favourite family member

- Eyes almost almond shaped
- Looking into space (dreamy eyes)
- Ears fall loosely to the side
- Mouth closed
- Lips relaxed

TIRED

Resting and content, with no need of anything

- Eyes noticeably almond shaped
- Eyes not looking at anything
- Eyebrow position neutral
- Ears hanging loosely to the side
- Hair close to the body around the face

EXTREME FEAR

Panic on first meeting a Potbellied pig

- Eyes very large and round
- Eyebrows raised
- Ears pulled back at the base
- Tips of the ears tight
- Mouth closed
- Upper lip tense
- Hairs around the mouth and nose raised

ANNOYANCE

Irritation caused by a pestering male dog

- Eyes rounded
- Eyebrows pulled together
- Ears flat and pulled back
- Direct eye contact, hard eye and tensed eyelids
- Mouth closed
- Vibrissae and nose hairs raised

STRESSED

Growing fear of the Potbellied pig

- Eyes very round
- Eyebrows pulled up
- Ears pulled towards the back
- Tips of the ears held tightly
- Mouth closed
- Bushy hair around the mouth and nose

2. THE FIVE F'S

When dogs are very tense they show an increase in calming signals, or signs of embarrassment, such as shaking off or scratching. If a dogs feels he is in real danger, he will usually resort to behaviours categorised as the five F's of fear.

THE FIVE F'S OF FEAR

By Lily Merklin, Certified Psychologist and Tellington TTouch instructor for horses

Flight or fight are the two responses to fear that are most commonly recognised. They are usually represented as opposites, because it is assumed that some animals respond to danger by fighting and others by fleeing. This statement is only true in as far as every living being has a preferences. The choice that is made depends on the species and breed, but also on the individual's experience from incidents in the past. Horses are categorised as flight animals that will flee when faced with danger, while dogs have more of a tendency to attack. i.e., fight.

Even though there are species-related preferences, each individual animal, and human, has the ability to flee and also to fight. But that's not enough. There are three other responses that come to light when a fear response is triggered and – along with flight and fight – all five responses start with the letter F:

- Flight
- Fight
- Freeze
- Faint
- Fooling around

Freezing is best described as being frozen in fear, and it applies equally to humans and animals. It is most commonly seen when something very scary happens, very suddenly and unexpectedly. In the human world, we often see this response on TV crime shows. In relation to dogs, a freeze will take the form of the dog stopping and remaining completely motionless, the hair on the top if his neck may stand up and he cannot be persuaded to move. Some people describe a frozen dog as being 'stubborn', but this is not the case. We know that if a dog is shaking and refusing food he is not being stubborn; he is frozen in fear.

If the dog continues to be frozen in fear, he may faint – the fourth of the Five F's. In addition dogs have a 'playing dead' reflex, which is between freezing and fainting.

The final F response – fooling around – is more commonly seen in animals and in children. When children are placed under too much stress, feel frightened or insecure, their behaviour may become completely paradoxical: losing concentration, playing with something or tugging at the mouth.

In relation to dogs, it is triggered when a dog is feeling insecure or is over-taxed with the demands that have been placed on him. The dog's response is very similar to that of a child. He will suddenly lose all concentration, become distracted and act very silly. He may find something to play with, or offer all sorts of tricks, which is why fooling around is also called flirting. He will also use his mouth a lot.

This focus on the mouth is particularly interesting, as the mouth is the most direct connection to the limbic system. This area of the brain sits like a hemline (limbus) around the beam that connects the right and left side of the brain. The limbic system does not just influence a number of vegetative functions, but also governs emotional reactions, such as anger, rage and affection, as well as learning and memory. The effects of the Mouth TTouch from Tellington method has proved to be highly beneficial.

FLIGHT

Being afraid and ready to flee is, from an evolutionary-biological point of view, a process that serves survival and ensures the future transmission of genes. So, in one, sense it makes sense to be afraid.

Most dogs will flee from danger as long as they think they have a chance of getting away and, as in this case, out-running the attacker. However, if the individual has had negative experiences with other dogs, he may attack, or feign an attack from a distance, under the banner: 'offence is the best defence'.

While playing, it is possible that a game of chase turns into flight for the dog in the lead, and hunt for the dog who is chasing.

Potentially a difficult situation can develop when the dogs involved don't know each other, and you cannot predict what is going to happen. It is your responsibility, as your dog's guardian, to observe the body language of both dogs carefully so you can interrupt a potential problem if needs be.

FIGHT

If a dog chooses to fight as a strategy, it is usually the people involved who have allowed the situation to escalate. Fights are usually about resources or lack of space.

Here are two males who are engaging in a fight. The group of Golden Retrievers approached the grey long-haired dog who showed too much interest in the females. With all the other dogs close by and no ability to leave the situation, there was no possibility for the grey dog to evade a fight and solve the issue peacefully.

The situation is very different for the dog on leash. Fleeing is impossible and the handlers are too close to each other to allow the dogs to feel safe. The dogs may also worry for the safety of their guardians, so the only choice they have is to attack.

FREEZE

Freezing can be triggered by a number of circumstances. Sounds and scents that we cannot perceive can cause a dog to freeze. Therefore it is even more important to recognise when a dog freezes, because the situation feels very threatening to him. It is often difficult to predict what he will do next. If your dog freezes, the best plan is to carefully attach a leash. This will make him feel more secure and help him to make a good decision, rather than bolt or attack.

Dhanyi, the Golden Retriever, has noticed something that nobody else is aware of. She stops suddenly and freezes for about 30 seconds. The photo below clearly shows how frightened she has been.

FAINT

This response is not seen so often, and guardians may not appreciate what is happening. Take the example of Gabi's male dog, Tiber.

Tiber had been poorly socialised by his breeder and, as he was growing up, he experienced panic attacks when he met strangers, or unfamiliar dogs. During his first 18 months, Gabi also realised that he had problems with his vision, which probably started when he was a puppy. His impaired vision caused him to develop huge insecurities. By age three he was completely blind due to a progressive atrophy of the retina.

With the benefit of wider experience, Gabi now understands that when Tiber 'fell asleep', it was a progression of freezing, which can turn into fainting.

Fainting can also be the result of undue stress and severe exhaustion. This is rarely seen in dogs. However, it has been witnessed in sheep herding trials where the sheep have fainted due to extreme stress.

Note the puppies in this photo. Each is wearing a harness for the first time and have collapsed into a faint-like state. The information that flooded their nervous systems, as a result of this experience, was too much for them and caused their bodies to respond in this way. Shortly after, they recovered and were happy to play, wearing their harnesses.

FOOLING AROUND/FLIRT

As already highlighted, fooling around is a behaviour that we often see in children as well a dogs. It is usually a sign that the youngster is overwhelmed with a given situation and can no longer concentrate. The response is to fool around and act silly, as a way of masking feelings of insecurity.

Dogs will often fool around with people when they don't know what else to do. Sometimes they will do the same towards other dogs and it looks like they are pretending to be puppies again. Fooling around is a polite plea for a break, a request for more mindfulness or a change in the situation.

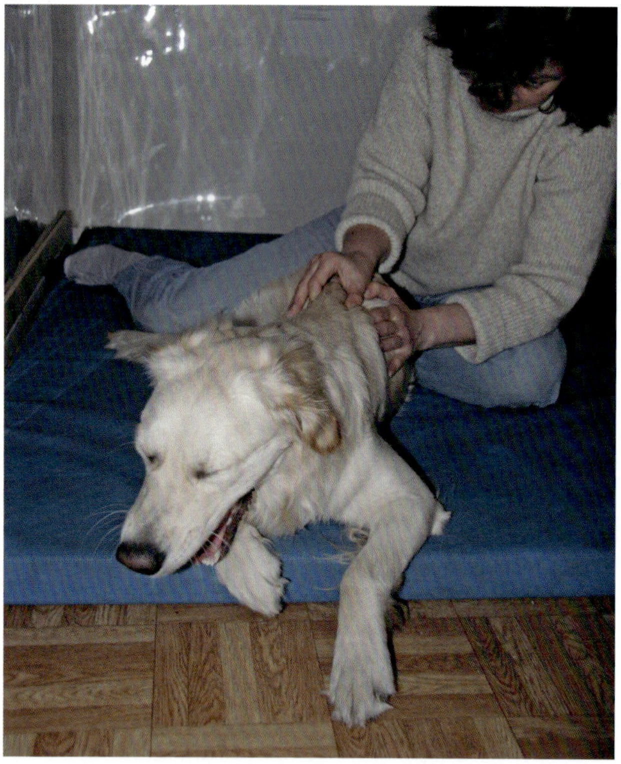

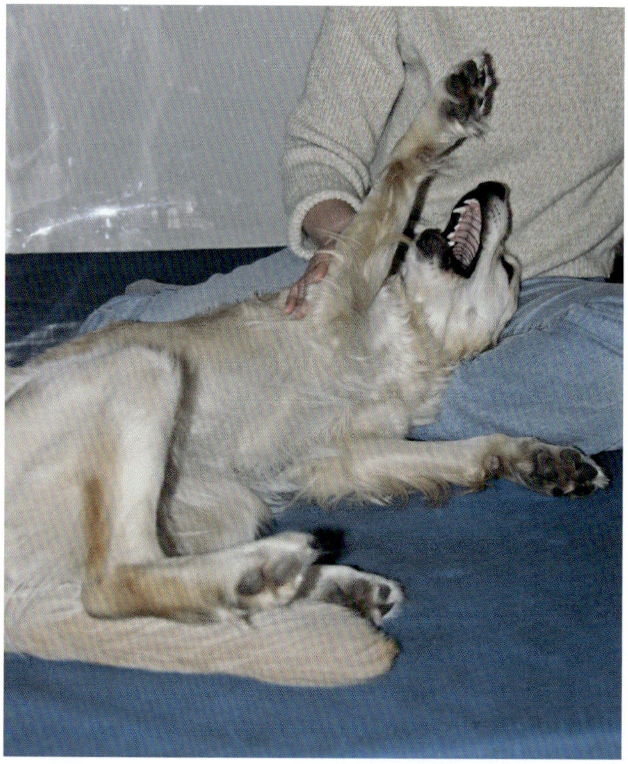

From puppyhood, Rani always struggled to find her balance (*see Shifting the balance point, page 226*) and she did not have good body awareness. Following an examination, the veterinarian could not diagnose a reason for this. We were able to help her find her balance with the aid of osteopathic and Tellington TTouch® treatments.

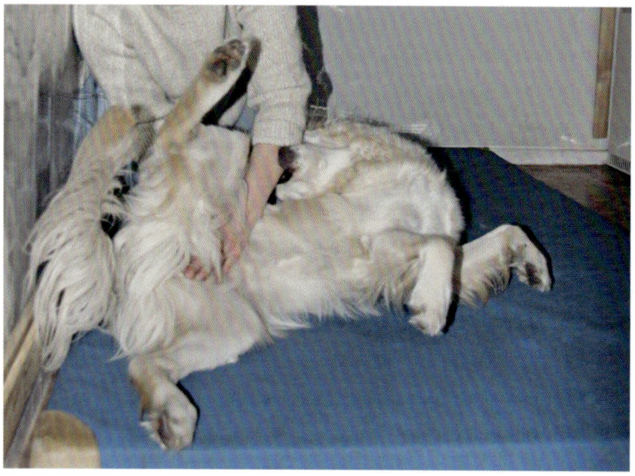

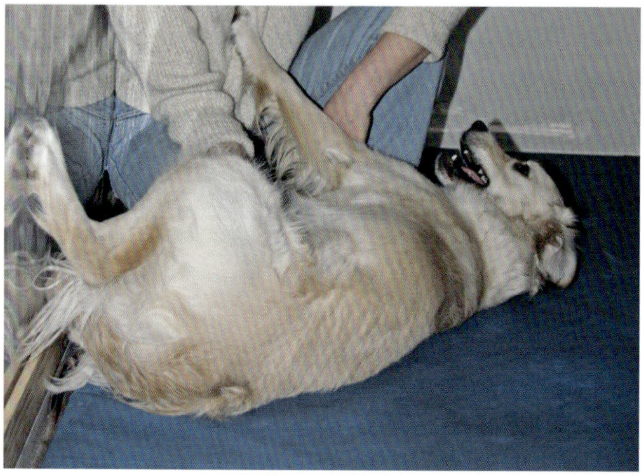

At the beginning of the osteopathic treatments Rani was always a little insecure, even though she loved contact with people. This series of photos shows her concern; she is fooling around in order to avoid treatment. We responded by asking the osteopath to pet her for a long time before starting the treatment.

3. ADDITIONAL SIGNS OF STRESS

Aside from the five F's, there are a number of other signals that a dog will send when he is experiencing stress. These signals are used often and may be stacked on top of each other; for example, a dog may use a number of communication signals at once. Other behaviours will be exaggerated, such as licking or barking. These behaviours reward the sympathetic nervous system of the brain, and should not be ignored.

A dog that is sending stress signals should always be helped. He needs to be taken seriously; his needs, fears and concerns are very real. Therefore, it is important to know how to identify these signals and to interpret them correctly, within their context. It is equally important to recognise when a dog has reached his limits of being able to concentrate, and when a situation is no longer bearable for him. Additional signs of stress to look for are: shivering, panting, excessive passing of faeces (diarrhoea) or urine, increased avoidance, attempts to empty the anal glands or very tensed muscles.

INCREASED USE OF CALMING SIGNALS

LICKING

Here, Quivive meets Katja's Potbellied pigs who live in her dog school and are used to meeting a lot of dogs and other animals. However, Quivive is not happy with the situation and would like to leave the enclosure as quickly as possible. Her increased and exaggerated use of licking shows that she is trying to calm herself. When communication signals are shown to excess, we know that the dog is under a lot of stress.

YAWNING

While under stress, dogs will also resort to using their embarrassment behaviour, such as yawning, more often. Here Lotti is yawning as someone approaches, a situation that always worries her.

BITING THE LEASH/PULLING ON THE LEASH

This behaviour is often seen when the dog is overwhelmed – for example, during a training session. Often the dog has co-operated and worked quite happily, and then suddenly grabs the leash. If the handler responds by pulling it back, it only makes the problem worse. Shorter training sessions, with more breaks, will help this type of dog.

Sometimes biting the leash is just a game – for example, when a puppy is getting bored walking on the leash. Retrievers like to carry the leash as they see it as their job so, in their case, it is not a stress signal. This underlines the importance of looking at the entire situation before making a judgement.

MOUNTING

Mounting is usually caused by a stressful situation and can be seen equally by males and females. Of course, mounting also has a sexual connotation so, again, we need to look at the entire situation.

Mounting can also be triggered by the scent of plugged anal glands or other physical ailments, such as a bladder infection. Dogs with these issue – and it does not matter whether they are male or female – are often pestered by male dogs.

Dhanyi mounts his step-mother Arwen. Dhanyi is two years old, and the behaviour has no sexual connotation. In fact, she is reassuring herself that she is still seen as a potential playmate. She will typically mount dogs that she has known since she was a puppy, as a way of showing she would like to play.

When in a stressful situation, a dog may respond by mounting or humping a person, usually on the leg of a trusted guardian. This serves to decrease growing stress levels, but it can also be a plea for help. When a dog mounts objects or toys, it is usually when he is trying to control his own excitement, for example, while greeting someone.

Quivive likes to stand up on Gabi and wrap her front legs around her guardian's leg. This is seen as a mild form of mounting or jumping up and, in the case of Quivive, it was her characteristic way of seeking help. In a relaxed situation, she also used this mounting to ask for a more intimate connection.

JUMPING UP

The act of jumping up helps the dog release tension. Tellington TTouch®, combined with positive reinforcement, can be beneficial. Of course, it is a personal decision as to whether this behaviour is tolerated, or not. In fact, there are some guardians who actively encourage it. It also appears that some dogs will only jump up on people when they receive this type of encouragement.

Jumping up on someone usually happens when the dog is over-excited. It may be because he is happy, or because he finds himself in a negative situation and feels stressed.

DESTRUCTIVE BEHAVIOUR

This behaviour is usually shown by dogs who have never learnt to be comfortable about being left alone. Dogs release stress through the mouth and, inevitably, the furniture, and other household objects, are targeted. When a young dog shows destructive behaviour, it can be simply that he is teething.

SHOWING THE PENIS

This can be seen in many males when they are excited and has no connection with sexuality.

EXCESSIVE BARKING, HOWLING

Barking, whining and howling are part of normal dog communication. Dogs can choose from a variety of barks, from a gentle woof to a loud bark, to express their different emotions.

Barking is self-rewarding behaviour and should not be ignored as it can become an anti-social habit. Dogs who bark excessively have something to say. It is up to us to listen and find out how we can help them interrupt the behaviour, and show them that they can communicate quietly.

There are some sounds, such as sirens or church bells, that trigger a dog to howl. Some males howl when they catch the scent of a female in heat, and some dogs will howl when there is a full moon. Many dogs howl when they are afraid of being left alone; they most definitely need our help.

SWEATING

Dogs mostly sweat through their pads. You can sometimes see sweaty paw prints when a dog is over-excited.

HAIR COAT AND DANDRUFF

When a dog is stressed, for example, during a visit to the vet, there may be evidence of hair shedding and dandruff.

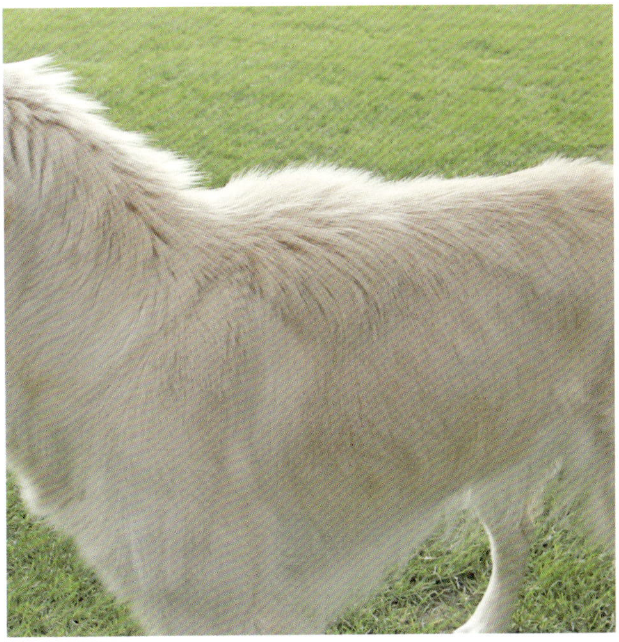

Raised hair

Raised hair, no matter where in the body, is connected to levels of excitement. (*See The Coat, page 300*).

CHANGES IN THE TONGUE

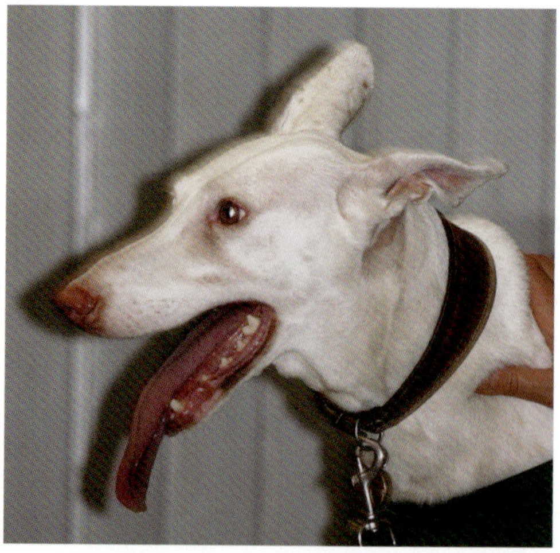

When a dog is very stressed, the tongue can change in different ways:

- Long tongue, hanging far from the mouth, widening at the tip
- Blueish shade due to lack of oxygen or heart problems
- Pale tongue – poor circulation
- Foam on the tongue

(*See The Tongue, page 150*)

DILATED PUPILS

While under stress the dog's pupils will usually enlarge as a result of their connection to the sympathetic nervous system (*see Iris and pupil, page 66*).

REFUSING FOOD

When a dog is showing signs of fear, including shivering and freezing, he will refuse food. This is a feeling we can relate to in times of extreme stress.

Here, Quivive and Gabi meet the Potbellied pigs for the first time. She was very frightened, and refused the treats she was being offered by turning her head away. In time, when she got used to the pigs, she was able to take the food.

Refusal of food is a good indicator of a dog's stress level. It means that he feels threatened existentially, and if he cannot resolve this situation himself, we need to step in to help and support him. Learning cannot take place without the presence of parasympathetic actions, something we need to be aware of when training our dogs.

When the dog can accept food again, he is giving us a sign that his stress level is lowering. When we offer food, it should be soft, smell good and be very small, because often it is not possible for a dog to chew in these situations. Swallowing food activates the parasympathetic nervous system which we urgently need in order to resolve difficult situations. Dogs that are naturally choosy about their food, will usually stop eating while under stress, much sooner than those who like their food.

There are times when a situation cannot be immediately rectified. In this instance, the Ear TTouch from the Tellington method is very helpful, especially when the dog has shut down and is failing to respond.

PARASYMPATHETIC AND SYMPATHETIC NERVOUS SYSTEM

All vertebrates have a vegetative nervous system that is able to help the body adapt to different life situations without conscious control of the brain (autonomously). It is divided into parasympathetic and sympathetic branches. The parasympathetic nervous system is responsible for the unconscious area, including breathing, heart rate and digestion. These are functions that are important for rest and recuperation. The sympathetic system regulates the range of instinctive reactions, i.e., it is responsible for fight and flight. It is activated when a life-threatening situation arises, or is perceived.

However, our brains are not able to distinguish between real and perceived dangers, and this is equally true for our dogs. Therefore, a situation that we do not perceive as frightening can trigger real fear in a dog due to his previous experiences, or lack thereof.

PUFFING OUT THE CHEEKS

This behaviour is most commonly seen in Sighthounds, Collies, and Dobermanns, and in associated crossbreds. Sometimes the dog will also enlarge his pupils while puffing out his cheeks

This subtle sign of stress is more prevalent in certain breeds. The photo on the left shows the dog's face in normal mode; the photo on the right shows puffed out cheeks.

DROOLING

Drooling, and excessive production of saliva, can be triggered through excitement and is often seen in dogs that are not very comfortable in the car. Drooling can also be a sign of illness.

When the Jacobson's organ (VNO) is activated – for example, when the male gets excited by the scent of a female in heat – you may observe lip smacking, the flehmen response and licking the air in addition to drooling (see *The Nose and Respiration*, page 168).

Foam around the lips occurs when a dog is highly excited.

Foam may form around the nostrils following excitement and exertion.

RESTLESSNESS

Restlessness, combined with stereotypical movements, is one of the disruptive behaviours that should be investigated by a veterinarian who specialises in behavioural issues

This female lives in a shelter where the dogs co-exist in groups and spend a lot of time outside. Although she would settle inside, she would run continuously when she was outside until she was completely exhausted. Therefore, her caregivers made sure she was not left outside for too long.

TAIL CHASING

Some dogs, especially youngsters, find tail chasing to be a fun game and will do it when they are in an exuberant mood. This game does not usually last long and is easy to interrupt. Some breeds, such as German Shepherds, have a genetic predisposition to tail chasing.

However, in certain circumstances, for example, on-going stressful situations, extreme itchiness, lack of socialisation, or stimulation, constant fear, or unsuitable/harsh training, tail chasing can become habitual and turn into a self-rewarding behaviour.

These dogs can also develop other obsessive behaviours, such as constant licking of their body, blanket or other objects, sucking, excessive barking, obsessive, mindless running in circles, or catching imaginary flies. These often start as coping mechanisms when there is a build-up of on-going stress, and will quickly turn into habits that the dog can no longer control. Dogs with these debilitating behaviours should be seen by a veterinarian who specialises in behavioural issues.

4. FEAR

Fear is a survival strategy. Those who don't know fear are at a much higher risk in a dangerous situation. Sometimes it is smarter to flee than to stay. Fear is a very healthy and life-preserving emotion and, when necessary, the body releases a lot of energy to be prepared for fight or flight. When fear turns into panic, or if a fear trauma is repeated, the animal's life energy or life force is compromised. This will weaken his immune system and can lead to devastating physical and emotional damage.

DEFINING FEAR

In our human world, the terms fear, fright, phobia and panic are defined as follows:

Fear: This is a feeling that is experienced by an individual in a dangerous situation, or perceived as dangerous, by that individual. The fear evaporates as soon as the danger has passed.

Phobia: This is a strong, on-going fear of certain things or experiences that are generally not dangerous – for example, a fear of spiders.

Panic: When fear is experienced so intensely that it appears without reason, and the individual experiencing the fear can no longer respond, we are talking about panic.

Since most dog guardians don't use theses specific terms, we have referred to 'fear' and 'insecurity' when describing the different facets of this emotion.

We need to be able to recognise fear in our dogs and to be able to respond so the dog trusts us, knowing we will take care of his needs and keep him safe. When a dog is afraid of fireworks, a thunderstorm, or something that has startled him, we should show understanding but not endorse his fear. It is not helpful to empathise to the extent of sharing his concerns, and suffering with him. We need to give support and help him to be less fearful, rather than pressuring him to be okay or pitying him. This is an opportunity to build trust and and develop a sense of security.

It is proven that young dogs that have negative experiences at a certain stage of their mental and physical development, can hang on to these fears for a lifetime. However, even adult dogs who experience a traumatic event can be affected for the rest of their lives.

Fears can have a lot of different causes. Besides traumatic experiences, pain, illness, and failing capabilities as the dog ages, can all result in fearful behaviour.

PICTURES OF FEAR

The physical expression of fear can be seen through posture, and the position of the ears, tail and legs.

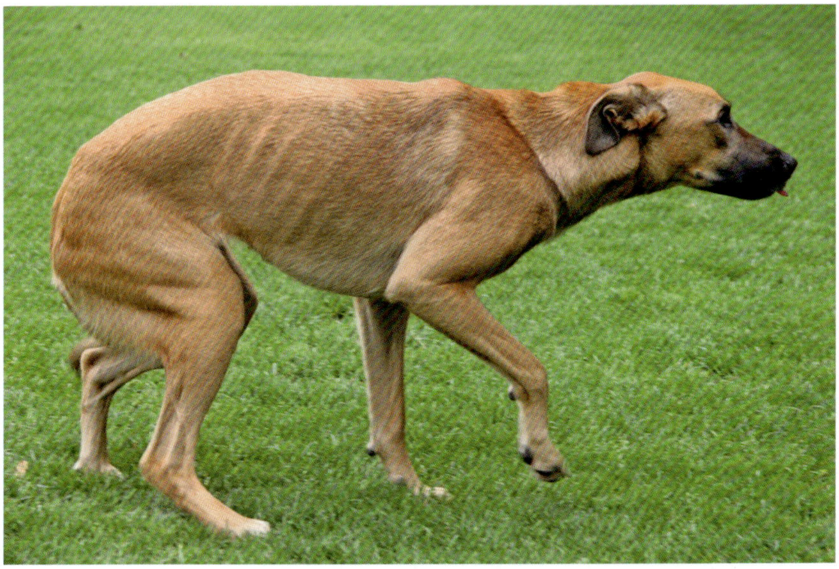

Lotti suffers from hypothyroidism that has never been treated. She is fine when she meets other dogs, but very fearful of people.

- Curved back, especially towards the tail end, tail tucked and not visible
- Neck and head carried below the level of the back
- Ears flat and pulled back
- Tongue flicked
- Joints of the hind legs bent
- Paw lifted

Compare the photo of Lotti to a Border Collie at work *(page 416)*. If you looked at this photo, without context, it would be easy to misinterpret his emotions.

The crouching posture, tucked tail, pulled-back ears, and open mouth with tongue hanging out, can give the impression that this dog is fearful. However, the raised hind end contradicts this theory. This dogs is working and fully concentrating on a herd of goats. The wide open mouth and hanging tongue are due to the exhaustion the dog is experiencing on a very hot day.

The dogs in both these photos are showing signs of fear: tensed backs, crouching, bent joints of the hind legs, tails either very low or clamped, ears flat against the head and pulled back, eyes large and round, wrinkles below the eyes and long mouths. The photo also shows exposed teeth. When afraid, many dogs will initially bare their teeth, and may bite if they feel that they can't get away.

FEAR IN YOUNG DOGS

A young dog may experience insecurity or feel afraid if he comes across an unfamiliar object, or an object that is familiar but seen in a different location. This may appear ridiculous, but the dog's fear is real. We need to remember that fear is a strategy for survival; it will help him manage dangerous situations, or those he perceives as dangerous. Our brains – both human and canine – cannot differentiate between a real danger, and a perceived danger.

Arwen's posture shows that she has seen something very spooky. She is pulling her ears back, her back is crooked, and she is crouching. Her point of balance is towards the back and left, which she achieves with minimal change in position.

Arwen is standing a little more upright, but her point of balance is still shifted back. She turns her head slowly…

…towards the rubbish bin, which has been moved to a new location.

FEAR OF AN UNKNOWN OBJECT

Adult dogs can also show strong fear responses when something suddenly changes in their familiar environment.

Rani and her daughter, Dhanyi, have come across some refuse bags int the forest. Rani (*lighter in colour*) approaches carefully with a lowered head. Dhanyi cannot decide if she should attack or flee. The hairs on her neck and back are raised, her hind legs are bent as if she is jumping while her front legs are stationary. Her ears are further forward than Rani's, and her tail is wagging vigorously.

MEETING OTHER DOGS

When dogs are well socialised, know their communication signals, and are polite towards one another, encounters are peaceful and harmonious.

Sadly, this is not always the case. One dog may be pushy and ignore the signals sent by the other dog. If the dog is insecure, this response can lead to fears when meeting other dogs in the future.

Chocolat is very concerned by a male Golden Retriever, who is pestering her with intrusive sniffing. She clamps her tail, her back is rounded and tucked and her knee and hip joints are folded. She has lowered her head and pulled her ears close to her body and back. Her mouth is pulled back, in what looks like a crooked smile. She tries to walk away from the male to escape his attention.

Here Chocolat shows a very different posture when she meets a male dog who approaches her politely and gives her space. Her back stays straight, her head is raised and her ears are in the flirting position.

WHEN THE GAME TURNS SERIOUS...

... or if it is a game for only one of the dogs. To assess a given situation, we need to observe the detailed body signals from both dogs. When size, weight and age don't match, the 'game' can easily develop into something more sinister.

Stevy (Bichon Frisé), is a calm and happy dog who likes to connect with other dogs. Here he is in a large, off-leash area surrounded by a number of dogs. The liver-and-white Spaniel puppy (in the foreground) wants to have some fun with Stevy.

The situation is getting a little too stressful for Stevy and he would like to create more space for himself.

While the adult dogs understand Stevy's signals and respond by turning away, the puppy misinterprets him and thinks he wants to play. This is a very common misunderstanding.

Stevy is turning into the 'prey' in this game of chase, which is not of his choosing. However, when dogs, like Stevy, have a lot of facial hair, it is not always easy to read their intentions. Small dogs often offer to 'fool around' to get out of situations such as this one. The people watching thought it was a fun game of chase for both dogs, but it is clear that the only one having fun is the Spaniel puppy.

Here Stevy is trying to get away from the Spaniel puppy and, accidentally, slams into a larger, light brown dog who just happens to be walking by. This could have caused a major misunderstanding between Stevy and the larger dog because, technically, Stevy's behaviour was very rude.

However, the confident, larger dog understands the situation and continues on his way without reacting. Stevy is usually a very polite dog who would not run into another dog intentionally. In this situation he was more concerned about his own safety and did not want to take his eyes off the puppy who was chasing him.

Now Stevy has taken flight and the excited puppy, who still sees the interaction as a fun game, is in hot pursuit. Stevy is barely fast enough to avoid being caught.

At this point, we asked the guardian of the Spaniel puppy to intervene and recall his dog. The size and weight difference between the two dogs was not a good match, and the Spaniel's sharp puppy teeth could easily hurt Stevy. When the dogs are moving this fast, it is not always easy to read their body language and postural changes. As the game was getting more stressful for Stevy, his movements became more hectic. Although his face and ears are difficult to read due to his heavy coat, his eyes are rounder and larger and his respiration flatter. His tail is no longer curled over the top of his back, but held to the side.

After a short break to catch his breath Stevy is feeling fine again; he is carrying his head and tail high.

FEAR OF SOMETHING UNFAMILIAR

Tristan, a Golden Retriever, suffers from a paralysis in his thoracic spineas – a side effect of bacterial meningitis, which leaves him with an unsteady gait. Tristan is always very excited when he meets another dog. The chocolate Labrador Retriever he is approaching is a little irritated at first.

The Labrador stops and stares at Tristan (not yet in the picture) who is approaching with a 'strange' wobbly gait. The Labrador shifts weight to the side and appears frozen. His eyes are almond shaped and his ears are flat towards the back. His eyebrows are slightly raised, which has created some wrinkling on his forehead. His tail is raised and it does not move. He does not seem to notice the small grey dog, Ella, who is sniffing him.

As Tristan comes closer, the Labrador looks directly at him. He has moved his ears forward and slightly raised them at the base. You can see some light tension around his upper lip. He has shifted his point of balance even more to the side and back.

The Labrador is almost frozen; only his tail is moving. He is carrying his tail below his back, and his pulled-back ears are even flatter. His eyes are getting bigger.

When Tristan gets quite close and looks directly at him, the Labrador quickly turns his head away. His eyes are still getting bigger and he shows a whale eye. He needs to keep an eye on this strange dog.

The Labrador shifts his point of balance even more to the side. His eyes are large and very round. He starts to move away.

The Labrador avoids Tristan, without making direct eye contact, and flicks his tongue. He shows relief at surviving this strange encounter.

FEAR OF STRANGERS

When a dog is insecure about meeting people he hasn't encountered before, it helps to make yourself small by getting down on one knee and avoiding direct eye contact, or even talking to the dog directly.

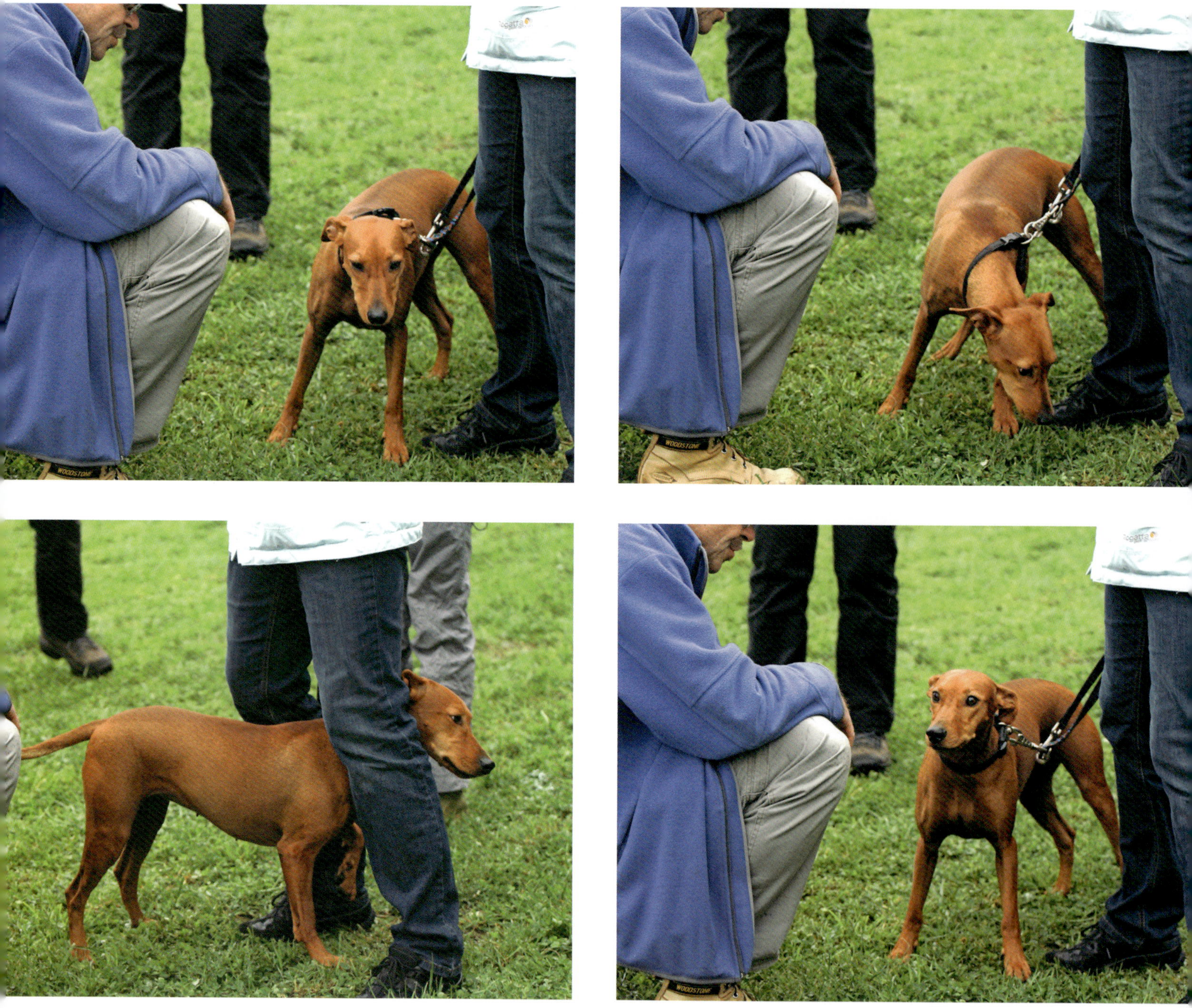

This meeting is causing the female Pinscher great concern. The man would like to make a connection with her and squats down. He has turned his body towards the dog and looks directly at her. However, he is not leaning into her space and is keeping his hands to himself. He may be talking to her. The Pinscher cannot handle the man's close proximity. She leans away from him, lowers her head as if she is about to sniff and looks away. She then resorts to hiding between her guardian's legs. When dogs are this fearful, and are on leash, they can easily bite.

UNKNOWN CAUSES

Emma, the black Labrador Retriever, was diagnosed with severe hip dysplasia when she was a young puppy and this has had a major impact on her behaviour.

Emma is disturbed by something that nobody, but her, is aware off. She could have heard something that the people with her missed, and the other dogs did not think was worthy of a reaction. She is very tense, which you can see by her high tail carriage, raised hair on her shoulders and stiff posture. Emma's tension increases. She starts barking and jumps forward.

Now she jumps back and to the side in panic. The hair on her shoulders is more noticeably raised.

If a dog is a so panicked he is no longer able to respond, you may need to put him on a leash to keep him safe. The next step is to find the source for his concerns and reactions. Pain, illness, or failing senses are more likely to be the cause rather than a traumatic experience.

5. SADNESS

Sadness has many faces; they are not always easy to read, or as obvious as in the case of the dog pictured here. He has been living in a shelter for a long time and, although there are many excellent establishments where dogs enjoy a good quality of life, they can never replace a forever home and the love of a special person.

DEFINING SADNESS

It is especially difficult for dogs who have spent most of their lives with a family, or a special person and then, all of the sudden, find themselves in a shelter. The death of a guardian or the break-up of a family is the reason why many dogs lose their homes. Aside from losing the person they loved, and their familiar environment and routine, these dogs have to deal with a new place and a completely different lifestyle. Older dogs are rarely chosen for adoption. In this situation, many dogs resign themselves to their fate, while others spend the rest of their life waiting for the return of their loved one. We may look into the eyes that look at us, but we don't really make eye contact. It is not always easy to see such sadness.

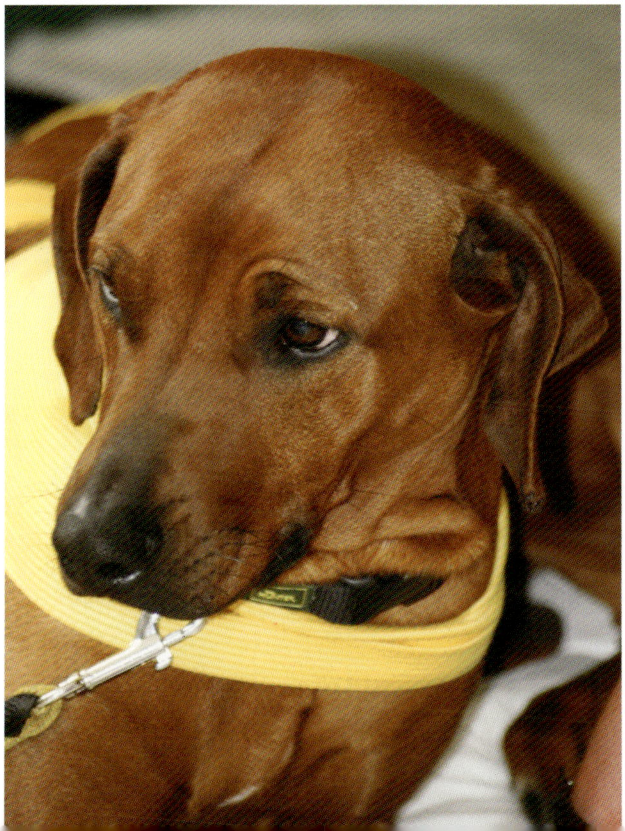

A STATE OF GRIEF

Dogs will grieve when other dogs in the family pass away. This female Ridgeback had to deal with the loss of both the dogs who lived in her family within a very short period of time. Everything she had loved to do before, she now did without joy. She acted like an old dog, even though she was only three years old. She showed very little interest in her surroundings, even when taken to new places. It was as if her life force had been turned off.

DEPRESSION

Female dogs can become very depressed if they experience a false pregnancy. Although it is not a tangible loss, the female concerned may be depressed, difficult to motivate and lethargic during this time.

Quivive is a good example of this. She participated in all her usual activities – clicker games and Tellington ground exercises – but she failed to show joy in what she doing. She would not take any treats, not even with the clicker. During this time, she carried her head lower than usual, and her tail – generally curled over her back – hung down loosely.

You can see the depression in Quivive's face: the ears, the long eyebrow hairs and the hairs around her mouth. Her eyes are bigger and rounder than usual. She looks into nothingness and fails to connect with her beloved family.

MOOD SWINGS

We rarely see sudden changes of mood in dogs. If it does occur, for no obvious reason, there could be an underlying physical issue. At the time, there was no apparent reason for the mood swing seen below but hypothyroidism, diagnosed at a later date, may have played a part.

Katja's deaf Border Collie, Momo, experienced huge mood swings. Here she is standing in a field, happy as can be. Her head and ears are raised, her tail is curled and there is a smile of her face.

Within seconds Momo collapses within herself. She lowers her head and ears, and clamps her tail. Her back is rounded back and she has lowered her hind end.

6. ANNOYANCE, ANGER, RAGE

There are differing interpretations of annoyance, anger and rage and, to some extent, they will always be subjective.

In brief:

Annoyance is characterised by irritation, frustration, and a degree of worry.

Anger is a powerful, often hostile response, usually directed at a provocateur.

Rage is violent and uncontrolled.

Our intention is to show the escalation from annoyance, to anger and, finally, to rage. However, transitions can be very fluid depending on the previous experiences or character of the dog.

WHAT IS ANNOYANCE?

"The annoying thing about getting annoyed is that you harm yourself without helping others." Kurt Tucholsky

When a dog experiences annoyance, he is usually feeling insecure, fearful, or frustrated, because his signals have been consistently ignored. When a dog comes too close to him, he is probably afraid of being pestered or getting attacked. A confident dog rarely gets annoyed, and very little bothers him. He is above all the things that upset less confident dogs. If the dog's annoyance also triggers his guardian's annoyance, the problem quickly spirals out of control. It is nearly impossible for the dog to find his way out of this situation without help.

A dog, on leash, may get very annoyed when he comes too close to another dog and, because of the leash, he is unable to take avoiding action. This behaviour may frustrate his guardian, if they feel the other dog presents no threat. In these situations, the dog's guardian needs to keep a cool head and take a deep breath instead of reacting to the dog's negative behaviour.. The dog should be offered an alternative response that shows him that we understand his problems and are concerned for his well-being.

This can be achieved by creating more space, for example, walking in a large arc or turning round and going in the opposite direction. However, if the dogs are already in close proximity, both guardians may need to intervene.

WHAT IS ANGER?

Anger is a very strong emotion and, at its extreme, a dog may be be 'blinded' by anger so he is no longer processing what he is doing. In confrontation, a display of anger may be so intimidating the other dog will not seek further engagement.

Anger can also erupt from a sense of injustice, when needs are not met, or when reprimanded for no understandable reason. Our dogs frequently find themselves in these situations.

WHAT IS RAGE?

Dogs usually show rage as an explosive and impulsive (reactive) emotion; they are often out of control and unable to respond when spoken to.

When a dog is showing rage, it is very possible that he wants to create fear in whoever he is raging against. Rage is not a rational response.

Rage can also be created by a situation that has triggered frustration, pain, or a circumstance that is perceived as unpleasant. Pain can cause a dog to become even more enraged.

We don't know exactly how dogs experience frustration, but we can imagine that it serves to trigger rage. Those with little self-control are more likely to experience this heightened emotion – something which applies equally to dogs and to humans

A dog that experiences rage repeatedly needs the help of a professional.

ANNOYANCE

Annoyance may be the emotion we identify most easily in the dogs we know well. Annoyance can be mistaken as insecurity, which is often the cause of it. We know this from dealing with insecure people who become annoyed when they feel over-taxed in a situation and don't know how else to cope. The transition from insecurity to annoyance may be very fluid.

ANNOYANCE BETWEEN DOGS WHO LIVE TOGETHER

Rani, the Golden Retriever, is quite annoyed/upset by the irritating behaviour of her housemate, Ella, who keeps trying to bite her hind legs, and is trying to make her stop. Rani is generally a very patient dog and always tries to de-escalate difficult situations. It is very rare that she shows annoyance this clearly. Her ears are pulled back and pinned flat against her head, the hair on her back is raised, and her dropped head and tensed mouth show she has had enough of Ella's pestering.

After releasing some of her frustration, Rani's face is relaxed again. Even though Ella is still barking at her, she ignores her and moves away, which is her more normal response.

Small versus large

When two dogs meet, and there is a difference in size, weight or age, we, as guardians, need to be especially mindful. We need to remember that a small or very lightweight dog can be seriously injured, or even killed, by a larger and heavier dog. To be kept safe, smaller dogs need our protection so they don't become emotionally scarred, or develop behavioural issues that are triggered by a sense of helplessness. Feeling helpless and vulnerable often leads to hysterical barking, escalating to rage, because these small dogs feel they are not seen or heard.

ANNOYANCE BETWEEN A YOUNGSTER AND AN ADULT

Rani's housemate, Tamika, could be really annoying when she was young. She did not always understand Rani's subtle signals, or maybe she wanted to test how far she could go... In general, it is better to prevent a smaller dog annoying a larger dog, even if it appears to be a game. The smaller dog is unlikely to hurt the larger dog but, if the interaction escalates, he may sustain serious injuries either intentionally or accidentally.

Here Tamika is biting on Rani's tail, which is one of her favorite games.

Now Rani is being pestered by Arwen who also wants to play. She turns towards Tamika, who is still hanging on to her tail, and is clearly annoyed.

Rani answers Arwen's request to play, which allows us to watch the interaction with less concern. We know that Rani can hold her own, indeed, she often initiates the games.

ANGER

Anger is an escalation of annoyance. It is very different from rage which erupts suddenly and can reach a point of no control.

ANGER DUE TO INJUSTICE

Tamika, the light-coloured female (*left*), bumps into Taschka in a provocative manner.

Taschka is obviously in a good mood and is not offended by the smaller dog bumping into her. She turns her head away and flicks her tongue.

This encourages Tamika to keep bumping her friend. Now Taschka's facial expression changes. She shows Tamika a whale eye and tenses her upper lip.

Tamika recognises the change in Taschka's mood just in time. She changes her approach by shifting her balance to the back and flicking her tongue to appease the larger dog.

Tamika, who is a peace-loving dog, feels she is being treated unfairly in this interaction. This situation could have ended in a serious fight, which is why it is not always advisable to let dogs work things out between themselves. For some dogs, this type of altercation could lead to a lifelong feud. It is best to recall the dogs, using a warm, encouraging tone of voice, at the very beginning of a conflict of this type. The dogs should then be kept apart. After everyone has cooled down, work can begin on restoring a peaceful co-existence in a controlled environment.

ANGER DUE TO PERSISTENT APPROACHES

Entire males can often be persistent in approaching other dogs. Here the guardians should call their male dog back because the other dog is very uncomfortable with the situation.

The grey female Berger, Verveine, keeps getting pestered by the black male, Laguiole. He just won't leave her alone.

Verveine tries to make it clear to the male that she wants to be left alone. However she does so with very little aggression because she respects the male with whom she lived when she was young.

Laguiole fails to respond to Verveine's non-violent signalling and, as a result, her patience comes to an end. She attacks him in anger.

Laguiole is able to jump out of the way and avoid the attack.

To support Verveine in this challenging situation, Laguiole was put on a leash until he was able to settle himself down.

ANGER BETWEEN A YOUNGSTER AND AN OLDER DOG

Many adult dogs dislike advances from younger dogs, and may react with anger.

Dhanyi, the Golden Retriever wants to play with Arwen, who is behind her, and reacts with anger when Neo wants to join the game.

ANGRY EYES

When a guardian has an intimate knowledge of their dog, they will pick up unusual or fleeting expressions. There were two occasions when Quivive, the Pyrenean Sheepdog, expressed her anger to Gabi.

The first time was when she was a youngster and, in boisterous mood, she attacked Tiber, who was old and blind. Quivive's guardian stepped in to separate the dogs. She held on to Quivive, and scolded her. Quivive looked at her guardian with angry eyes – an expression her guardian had never seen before . As a young dog, Quivive had no understanding of what was expected of her. She saw the interference and reprimand as an unprovoked, and unfair, attack.

The second time Quivive showed anger was towards a male dog. When he first approached, he showed his neutral intentions and she responded by walking around him in a large arc. After both dogs had passed each other and had already covered some distance, the male turned round and attacked Quivive from behind. She turned round in a flash and defended herself. She was full of indignation and her eyes sparkled with anger.

A VERY SPECIAL MEETING

Katja Krauss

Meeting the street dogs in the Arabian Emirates counts as one of the most fascinating experiences I have ever had with dogs. Therefore I would quickly like to share a special story.

I was there to teach a group of dog trainers, who were working for a sheik, but every morning I found time to take photos of the street dogs. After a few days of observing them, I got to know some of them as individuals.

One morning I parked beside a busy construction site. A tan male was inside the fence and only came out after a group of white dogs had spotted me.

It looked like the group of white dogs consisted of two females and their half-grown pups. I am sure all the dogs knew each other very well. It seemed as though the tan male was keeping an eye on the situation from a distance. The two females ignored him.

I was busy taking photos of the tan male, who was standing on the left side of my car, when all of a sudden he shot towards one of the white females just as she was about to attack me. When he briefly turned away, the female attacked me again.

What happened next really amazed me. The tan male faced the female and attacked her again. She fled far behind the car.

In retrospect, I believe the tan dog made a decision to protect me. I was a complete stranger yet he was willing to protect me even though he risked being injured himself.

The tan male stops an attack by this white female

The second time the male takes issue, the female takes flight.

ANGER DUE TO PAIN

Justified anger can be shown by dogs who don't want to be touched or bothered because they are experiencing pain. Guardians who have male dogs in their care need to prevent their dogs from mounting other dogs; it is generally seen as being very impolite behaviour. An older dog can experience a lot of pain when another dog mounts him and, in most cases, he cannot defend himself

The male Golden Retriever, Ronnie, is being bothered by a young Husky who is mounting him. Ronnie is getting on in years, and suffers from various ailments. He shows the Husky his whale eye, and you can just see the tip of the Husky's tongue. Ronnie's face is tense; his ears are pulled back, the top of his head is tight and his mouth is closed. He is trying to stop the Husky by making eye contact, but does not want to escalate the situation.

The Husky is undeterred by Ronnie's eye contact. Ronnie turns his head away from the Husky and raises it. The eye that is turning towards the Husky is getting larger. He pulls his left eyebrow up, a sign that he is getting quite annoyed.

Now his annoyance has turned to anger. He turns quickly towards the Husky and uses his teeth.

Ronnie's ears are raised and pointed forward at the base. His teeth are still clearly exposed, his whiskers are pointing forward. The eyebrow that is visible is still raised on the inside. Ronnie is deadly serious. His hind end almost collapsed under the Husky's weight, and he is most likely in a lot of pain.

Ronnie is having trouble stabilisng his hind end. Looking at the close-ups, you can see he continues his eye contact with the Husky, his whiskers are still standing up and his ears are forward. He does not want to be approached again. The Husky jumps past him and looks away from Ronnie. Due to his light-coloured eyes it is harder to see his whale eye.

RAGE

Physical ailments and emotional traumas are common in shelter dogs and can be triggers for sudden outbursts of rage.

RAGE AGAINST RIVALS

This black-and-tan male lives in a shelter and, because of his good social skills, he lives in a group with several other dogs. Behind the wooden fence is a pen with a Dobermann, who does not get along with other dogs. The black-and-tan male is enraged by the Doberman, even though he cannot see him. Being able to hear him, and smell him, is enough to send him into a rage. While other dogs in the group become involved momentarily, and then leave, the black-and-tan male becomes increasingly enraged. These two dogs have never seen each other but an animosity has grown up between them

The black-and-tan dog only settles after the Dobermann has been moved to another pen. The caregivers at the shelter are usually very mindful never to have the two dogs outside at the same time to avoid an incident such as this.

RAGE BETWEEN A YOUNGSTER AND AN ADULT

While there are two dogs engaged in some roughhousing in the foreground, more action is about to ensue behind them. The young Bernese Mountain Dog, who appears in the top right of the photo, has just come out of the water.

The young Bernese Mountain Dog is galloping at full speed towards the retaining wall. His posture is open; he is carrying his head and tail higher than his back and he is having a fun time.

He stops abruptly – and you can see why. A female Rhodesian Ridgeback (*arrowed*) is approaching him very aggressively. The Bernese Mountain Dog makes himself very small; you can hardly see him in the photo because he is blocked by the two dogs in the foreground who are continuing their wrestling match. It looks like the young dog is trying to change direction.

The Bernese Mountain Dog tries to flee. His head is low, his back is rounded and sloping down at the tail end, and you cannot see his tail.

The corner of the Bernese Mountain Dog's mouth is pulled upwards; his ears are flattened to his head and held to the side and back. The eye we can see is red, probably due to an irritated conjunctiva. Because of his lowered head and his attempt to see what is happening beside and behind him, his eyeball looks distorted. He is uncomfortable in this situation, which he did not create intentionally.

The female Ridgeback chases him. Note his clamped tail, and the iris that is turned towards the camera.

The female slams into Bernese Mountain Dog without hesitation and knocks him to the ground. She is in a rage and her behaviour is very rough.

When a young dog is disciplined by an adult dog, with this degree of severity, he risks serious injury. His skeleton, tendons, ligaments and joints are not fully developed, which makes him especially vulnerable. In addition, we should not underestimate the psychological damage inflicted on a young dog, which may only become apparent at a later date, usually after puberty. Even if there is no physical injury, a sensitive dog may develop life-long problems in the shape of fear, or aggression towards other dogs. These feelings could be triggered by dogs in general, or only by dogs of that specific breed. Whenever possible, the adult dog should be recalled in order to put a stop to such unfortunate experiences.

The Ridgeback seems to be able to control herself. She is shifting her weight away from the Bernese Mountain Dog and blinks. Her posture is still very stiff and her tail is held high. The young Bernese is back on his feet. He shakes either the sand or the stress – probably a combination of both – from his body.

The Ridgeback avoids direct eye contact, but her posture remains unchanged. Her tail is now slightly lowered. It still takes quite a while until the young Bernese Mountain dog dares to move, and slowly walk away.

Why such rage?

We don't know what triggered the female Ridgeback to be so aggressive with the young Bernese Mountain Dog. When we looked through the photos we had taken at the beach that day, we noticed that she kept a distance from other dogs. However, when she approached another dog, she was very boisterous. When she met a male Ridgeback, she asked him to play, but then did not follow through. It was a hot day and there were a lot of dogs near the water. Maybe she had already met too many new dogs that day and could not handle any more. This may have caused her to over-react.

Maybe she had a particular dislike of Bernese Mountain Dogs, or maybe she wanted to stop the young dog and take control of the situation. Maybe she misunderstood his intentions and felt threatened. Sadly she was not recalled until the young dog was on the ground.

FREE FOR ALL?

Many adult dogs like to discipline young dogs. The success of this strategy depends on the adult's level of experience; some may be too rough with the more vulnerable youngsters. Pain, physical limitations or diseases of internal organs, such as subclinical hypothyroidism, can lead to such impulsive actions. An off-leash setting, where dogs gain valuable experiences, can be enriching. However, it is important to also give your dog a break to assimilate the many impressions he has gathered.

RAGE AGAINST THE REFUSE COLLECTOR

Visitors to the house, such as refuse collectors, postmen and delivery couriers, are a common trigger causing many dogs to fly into a rage. These scenarios should be introduced, and managed, while the dog is still young.

This fit of rage was triggered by refuse collectors who were emptying bins on the other side of the fence. We don't know why this female dog had such a furious reaction. Maybe it was because of a bad experience in the past.

- Back hairs raised in the shoulder and tail area
- Head and neck stretched forward
- Ears are tipped far forward at the base
- Looking straight ahead
- Severe wrinkling on top of the head
- Mouth open and shortened
- Wrinkled neck
- Whiskers pointing forward

The enraged dog not only jumps at the fence and barks ferociously, she also retreats from the attack, and then looks very insecure.

Additional signs:

- Hair clearly raised on the shoulder and tail
- Mouth even shorter

Note that you can only see her top teeth when she raises her head high, and gravity pulls her lips back. This gives us the impression that she may not really be able to lift her lips.

Additional signs:

- Hair raised over the entire back
- Whale eye
- Upper teeth visible

She can only relax when the refuse lorry has left.

AVOIDING THE TRIGGER

There are many reasons for fits of rage and it is very rare that this intense emotion is experienced without cause. If there is specific trigger, you will need to manage your dog's environment to avoid such escalation.

Taschka carries a toy which helps her impulse control when dealing with other dogs. Without the toy, this interaction with the Cocker Spaniel would have played out quite differently.

Taschka:

- Tail raised high
- Ears pulled back
- Whale eye
- Leaning away from the Cocker Spaniel

Dogs release tension through their mouth. Carrying a toy activates the parasympathetic (vegetative) nervous system which is responsible for functions such as sleeping, eating and digesting food. To simplify, it is the area that has a positive effect on the dog's emotions.

Taschka never injured another dog, but she could scare a more sensitive individual. As a result, her guardian encouraged her to carry a toy in her mouth to curb her more impulsive behaviour. The cause for her behaviour was an untreated case of sub-clinical hypothyroidism, which remained undiagnosed for some time because the bloodwork was normal.

Sub-clinical hypothyroidism not only affects the physical wellbeing of the dog, it also results in numerous behavioural changes. In Taschka's case, she showed an increase in impulsive and reactive behaviour towards other dogs. This changed once she received mediation that stabilised her thyroid. However, in these cases, it may be necessary to teach a dog an alternative behaviour to break the pattern.

7. REVULSION/DISGUST

Revulsion or disgust is a primeval reaction that protects us from parasites, bacteria and other organisms that can make us ill. We want nothing to do with disgusting things and do all we can to keep them out of, and away from, our body. Our expression reflecting disgust is universal and spans all cultures: a wrinkled-up nose, eyelids pulled together, and upper lip wrinkled. This may be accompanied by gagging, salivating, and possibly vomiting.

What triggers our revulsion is dependent on our own culture. For example, boiled sheep eyes are considered a delicacy in the Middle Eastern countries, but most Europeans and North Americans would shudder at the thought of eating them.

Dogs also show disgust but, compared to us, it is not a major issue. We are therefore including just two examples. It is interesting to note that dogs use the same facial expression as us when they find something they find disgusting, including the same parasympathetic responses, such as drooling, gagging and vomiting.

REVULSION TOWARDS TICKS

Dogs express a strong sense of disgust when confronted with a tick. This was caught on camera when Nelly, a Rhodesian Ridgeback cross, found a blood-filled tick on the ground. Her guardian discovered that she is repulsed by looking at engorged ticks after they have been pulled off her.

We would like to thank (and apologise to) Nelly, who so very bravely cooperated with the photo shoot to allow us to show disgust and repulsion. She truly earned the many treats she received for her services!

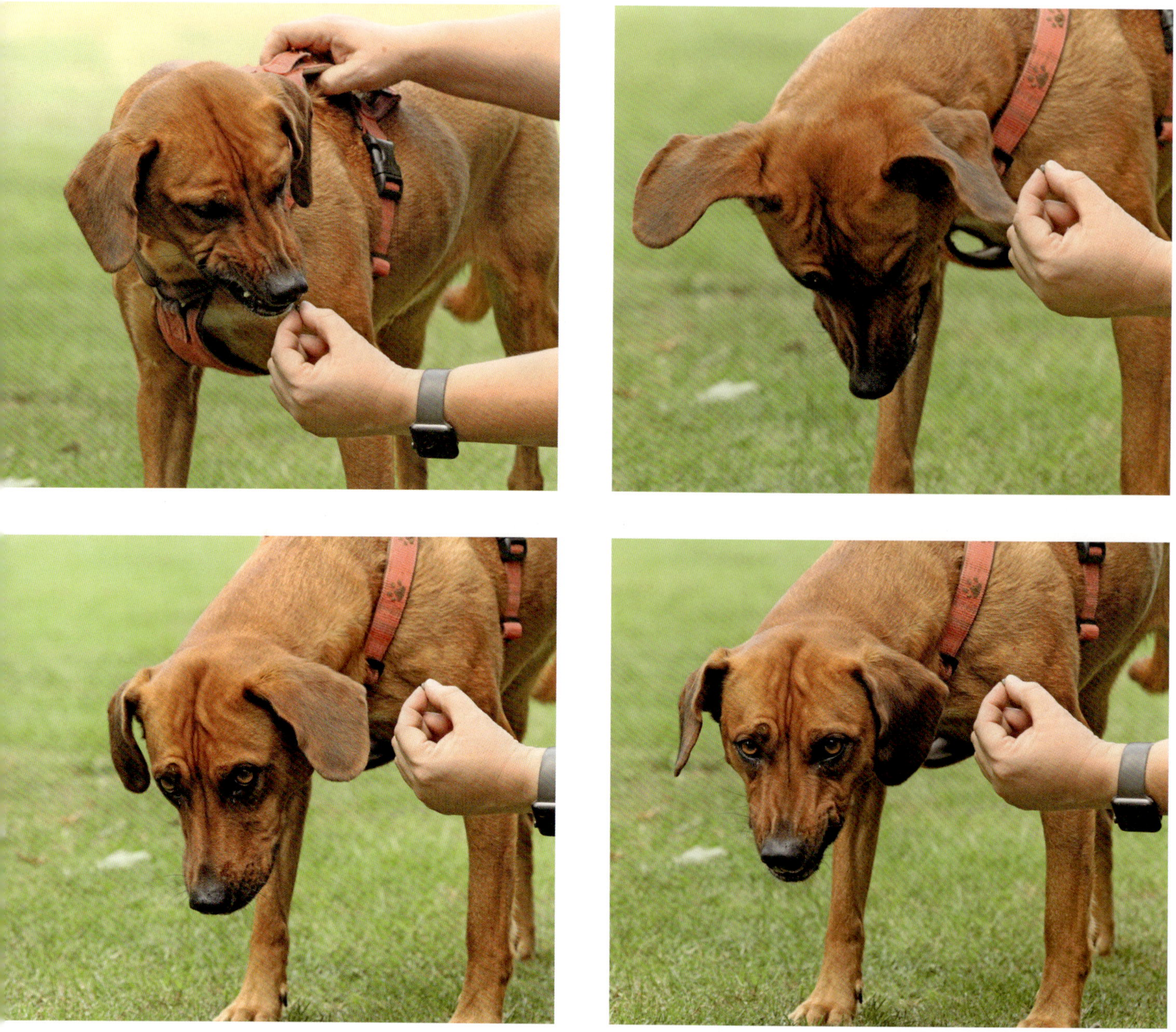

The expression of disgust and revulsion is not just inter cultural, but also crosses the species. Nelly's nose is wrinkled, her eyes squeezed tight and her upper lip is wrinkled.

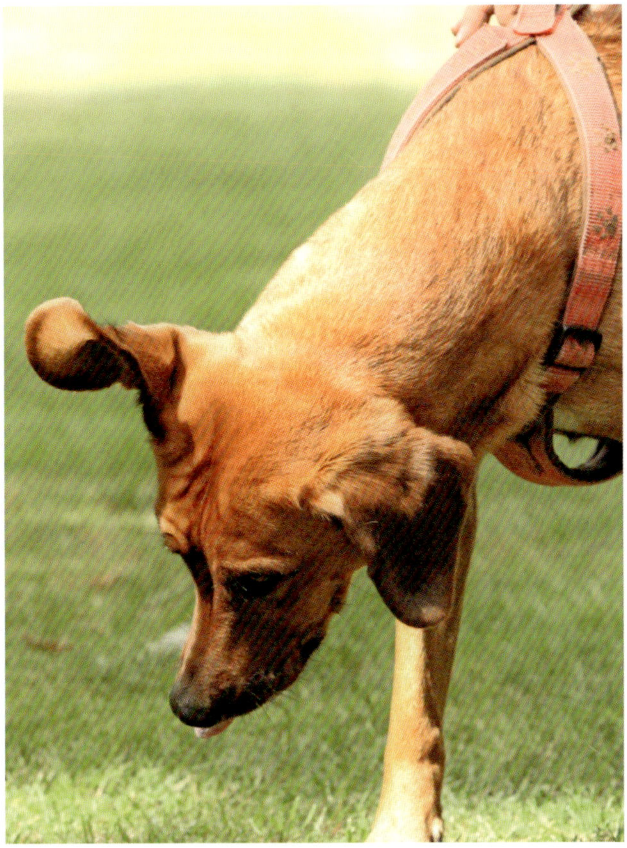

She shakes her head with disgust and starts to drool.

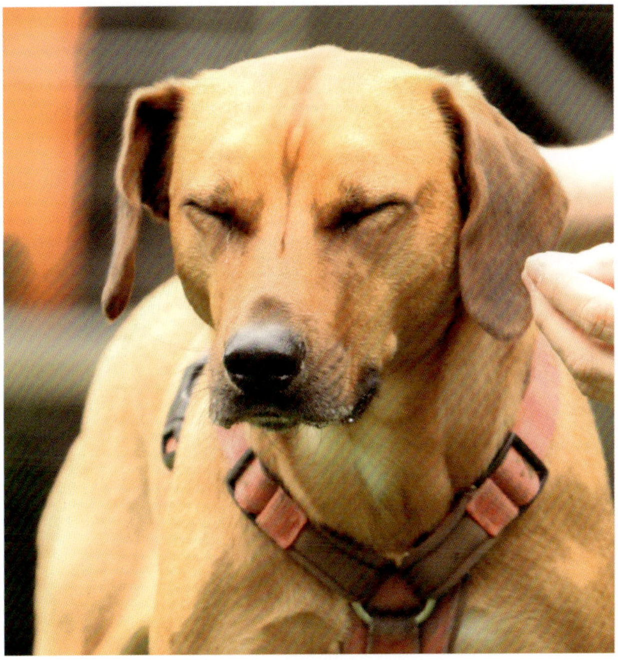

You can clearly see Nelly's revulsion when she is shown the tick.

REVULSION RELATING TO SURFACES

Dogs' feet are very sensitive and many will show an active dislike/revulsion when their feet come into contact with a particular surfaces or object. Needless to say, they use their sense of smell to examine the offending object.

This Vizsla is being led through the Tellington labyrinth (an obstacle used to teach leading with awareness). All of a sudden he stops because he noticed something on the ground.

Even though his guardian is giving him a signal to walk on, the Vizsla lifts his left front paw to examine the area more closely.

After a thorough examination he continues cautiously, making sure not to step on the offending spot. On that training day several other dogs showed the same hesitancy when they reached this corner of the labyrinth.

We have a lot of anecdotal evidence of dogs showing a particular aversion, which they express as disgust or revulsion. Take the example of Quivive, the Pyrenean Sheepdog. She had a particular dislike of sitting on the edges of trails where other dogs had relieved themselves. If she was asked to sit, she would hover just above the ground. Once her guardian, understood her sense of disgust, she simply asked her to stand and wait, when required, and Quivive was happy to co-operate.

Quivive also hated stepping in dog faeces – even her own – and would avoid any excrement she came across by taking a huge leap if she could not walk around it. As she aged, and her senses diminished, she became less sensitive to this issue. When her hind legs grew weak, and it became difficult for her to squat while urinating and, on walks, she would often end up standing in a puddle of her own urine. However, on returning home, she would lead her guardian to the shower to wash off the urine. This was especially noteworthy as Quivive had never been a fan of water and bathtimes.

8. JOY

We believe that joy is the easiest emotion to recognise and interpret in dogs. Just like us, dogs have many different reasons for feeling joyful.

THE JOY OF PLAYING WITH AN OBJECT

Here Rani, the Golden Retriever, expresses her joy as she plays with a shoe she has found.

THE JOY OF PLAYING WITH OTHER DOGS

A wild wrestling game is the best fun when both dogs are mindful of the rules. It always helps when both dogs are evenly matched in size, strength and age.

THE JOY OF PLAYING TUG-OF-WAR

This is a game that gives great pleasure to both dogs. However, a game of this type should be supervised to ensure it does not escalate and develop into a fight.

THE JOY OF PLAYING WITH A FRIEND

Some dogs are more reserved, or selective, about playing and will be happier engaging with a trusted individual.

THE JOY OF PLAYING ALONE

This dog was having a great time playing by himself, ending his game with a joyful roll. Rolling in grass, sand or snow is something that most dogs enjoy.

KNOW YOUR DOG

To ensure that playing is fun for your dog, you need discover his favourite game. Originally dogs were designed to pursue prey and grab it. For many dogs, running, carrying a 'trophy' and water are key ingredients of happiness. This can vary between individuals and between breeds, where certain preferences, such as retrieving, are genetically predisposed,

If two or more dogs play together they should be matched in age, weight and game preference (wrestling or running games). As responsible and loving guardians, we need to gently interrupt when one of the players is telling us that the game is getting out of control, or that one of the dogs needs a break. Otherwise the game can quickly change to a more serious interaction; at the very least only one of the two dogs playing is having fun (*see Fear, page 568*).

THE JOY OF DIGGING

You should never allow your dog to dig in fields that have been seeded, or pastures used for grazing, as the resulting holes can damage farm machinery or cause injury to people or animals who may step into them.

Digging in sand or soil can be a true passion for some dogs.

THE JOY OF WATER

Many dogs find great joy in water. They may display their joy by excessive barking or crazy games running in and out of the water. This is another situation where we need to decide if the dog is too stimulated and if a slower, calmer approach, or a break, would serve better.

THE JOY OF RUNNING

How much a dog loves to run is very individual, and greatly depends on breed. Some dogs opt for long distances, while others prefer a short sprint to stay closer to their people. Generally dogs have been born to run and it a very important activity for both their mental and physical wellbeing.

LAUGHING AND SMILING

Most dogs can laugh, while smiling seems to be linked to a genetic component. It can be seen in the offspring of certain lines, but not in others (*see The Mouth, page 78*).

Rani, as a puppy, with a big smile on her face.

Here you can see she retains her smile into adulthood.

This smile, where teeth are exposed, may be misinterpreted as a threat if you don't know the individual dog. Here Eve is making direct eye contact and the corners of her lips are pulled back, as they would be if she was threatening. In addition, it looks as though her ears are pulled back, something we would also observe in a dog who is threatening. However, Eve's tongue is loose inside her mouth and that tells us that she is not tense, and is enjoying a great time. Of course, Katja knows Eve is a great smiler!

HEARTFELT JOY

When connecting with a loved guardian

A dog may have a really strong bond with his special guardian, but it may still be difficult to make direct eye contact or accept a very intimate touch.

India is happy to interact closely with her guardian. The expression of her eyes, the direct eye contact, the positioning of her ears, and her entire facial expression, remain unchanged when her guardian cups her hands around her neck. Looking at the corners of India's mouth, it looks as though her expression has changed to a huge smile. This is not the case for many dogs, especially when contact comes straight from the front.

When loving and trustful connections are made while the dog is still a puppy, they will last a lifetime.

THE JOY OF GREETING

Dogs will often shake off after they have jumped up at someone to greet them.

It is a matter of personal preference as to whether you allow your dog to jump up during a greeting. If you don't mind, be sure to warn others who may be surprised by such exuberance, or children who may be overwhelmed.

GESTURES OF BELONGING

Towards people

Sometimes joy is not easy to see when relying on the dog's body language. There are those who would consider this dog to be showing submissive behaviour. However, it is a gesture of belonging, which is shown between trusted members of a family. It is used to strengthen that special bond, especially after a separation.

GESTURES OF BELONGING

Towards dogs

Some dogs show a very similar type of greeting (which also includes lying down and crouching) towards other dogs, which are easily misinterpreted as submissive.

Little Sally is very excited to see her next-door neighbour, Olli, who is an impressive Irish Wolfhound. She is not shy about jumping up on him during her greeting.

Here you can see how both dogs communicate with each other. Following her greeting, Sally lies down in front of Olli and lets him sniff her. Her eyes are their usual size and not overly rounded. Olli shows a whale eyes and licks.

SHOWING LOVE

This puppy is expressing her love for an older dog, which is not uncommon among youngsters. The puppy, Dhanyi, closes her eyes and looks as though she would like to climb inside Arwen. She loves her playmate dearly and still looks up to her to this day.

THE JOY OF PLAYING WITH CHILDREN

It is certainly not a given that dogs are relaxed and happy when they are around small children. There are some individuals who adore children, but the majority of dogs are worried and tense about interacting with them, possibly due to bad experiences or lack of experience. The dog will often take his cues from his guardian, and the children's parents, so their responses are key.

Here the young Golden Retriever, Kalea, meets her guardian's friend's young daughter. The two of them have met before, but they don't see each other very often. Kalea is very excited; she approaches the little girl with care, and lies on her back.

Terrier mix, Tamika, happily rolls on to her back in front of a seven-month-old toddler whom she knows well. Both dog and child are comfortable with the situation and enjoying the interaction.

JOY OF LIFE

Despite a handicap

A handicapped dog can experience as much joy in life as a healthy dog. If the guardian adopts a positive attitude, this will play an important role in the dog's ability to be happy (*see The tail, page 194*).

As the result of an accident, Luke suffered a femoral fracture and is paralysed in his right hind leg. Fortunately, it does not affect his ability to swim or run with the other dogs.

Luke sets a wonderful example of living a joyful life – something we could all learn from.

TRISTAN: A CASE STUDY

After contracting bacterial meningitis, Tristan was paralysed in the middle of his body. His veterinarians said there was no chance of him ever walking again. However, Katja and her family refused to give up hope. Firstly, the Tellington method gave him a better sense of his body and helped to restore his strength. Then a variety of therapies – homeopathic, acupuncture and osteopathic –were employed so that, over time, Tristan was able to walk again. *For more information about Tristan, see Pain and ailments, page 245.*

Today he accompanies his family on their walks and loves to play with his close friend, Dhanyi.

With the aid of a life vest, Tristan gets additional exercise and thoroughly enjoys his time in the water.

9. CURIOSITY

CURIOSITY IN PUPPIES

Puppies show curiosity, and a desire to investigate, when they are still very young.

Puppies should be exposed to different environmental stimuli while they are still with their breeder. Starting at about four weeks of age, they should begin to explore their environment, which should be set up for maximum enrichment. This helps to develop the use of their senses, and they will also become adept at using their paws, and their mouth, to investigate. These skills are very important and can be enhanced with some simple clicker training exercises.

CURIOSITY ABOUT WATER

Dogs who are exposed to water as puppies usually develop a great love for playing in it. However, there are those who dislike water, and avoid getting their paws wet at all costs.

There are two strategies: approach with care...

... or dive right in.

CURIOSITY PAIRED WITH CAUTION

Curious dogs have a strong desire to learn and are usually easier to train because they are open to new experiences. If the dog is also cautious, it is important to teach him in small steps and to ensure the experience is positive. Such ambivalent behaviour – veering between curiosity and caution – can be seen in a lot of dogs. The opposing emotions hold equal sway on the situation

Inca, the female Dobermann is preparing to meet Chemsi, a Sloughi puppy, who is about to become part of her family. During the drive to her new home, Chemsi was lying on the red sweater, which is now being presented to Inca. Cautiously, but showing curiosity, Inca checks out her scent. She is shifting her weight to the back, and her neck is stretched far forward towards the sweater.

Inca uses a similar posture when she first meets the new puppy. Inca is prepared to flee and Chemsi turns her back towards the older dog to calm her.

HEAD CARRIAGE WHILE CURIOUS

Most dogs will tip or tilt their head when they hear an unfamiliar sound or find themselves in a curious situation.

Inca shows her curiosity as she watches Chemsi at play.

Lavjama heard a kissing sound.

Ischka (*left*) and Arwen (*right*) both show their curiosity by tipping their heads.

Did Quivive hear a mole digging?

Some dogs lift a front paw, similar to pointing, when they have noticed something that awakens their curiosity.

CLOSING WORDS

It has been such a pleasure for us to gather all these examples to show you the body language and communication signals that dogs use to express their emotions. We hope that you will now see dogs with different eyes and be able to evaluate their behaviours more easily.

Please follow our suggestion and make a study of the photos. This will help you to develop an awareness of those small changes, and subtle signals, that our dogs use to communicate. It is not necessary to set up situations that could provoke issues, become dangerous or create fear. It is in those daily interactions where the subtle signals are used, and will elicit the most dramatic changes, if we can recognise them. The goal of this book is for you to ridentify that crucial moment before escalation.

From now on, you will probably look at dogs in the same way as we do, and become fascinated by what you discover. There is so much to learn about the emotional life of our dogs, and how much this, in turn, influences our ability to respond to their needs. If we start reading the signs, our dog will be encouraged to communicate with us on an even deeper level.

If you are focused on watching for all those small signs – and ask yourself why you had not spotted them earlier – then this book has achieved its goal.

THE AUTHORS

KATJA KRAUSS

Why I wanted to write this book

I am convinced that I must have been a dog in a previous life. Why else are the actions and reactions of animals so obvious to me while, unfortunately, so many people seem to be unable to perceive them?

From a young age, you could always find me surrounded by animals – baby birds, cats, mice – and a stuffed brown bear, who always wore a collar and leash, and accompanied me everywhere. I chatted with the animals I met, and I soon got involved with training our three family dogs. I always worked from the dog's perspective, trying to work out what each individual was feeling and thinking. Back then I called it "translating", verbalising what the dogs were thinking, or, rather, trying to interpret their emotions from their behaviour.

We have the opinions; the dogs have the facts. I would never assume that I know exactly what is happening inside an individual but confirmation comes in the form of the reaction that follows an interaction. So far, I have had 17 of my own dogs join me in my life's journey and I have come into contact with thousands of dogs in my training school. Therefore I speak from experience and I follow my gut instincts. When I describe something in this book, it is my opinion and I don't claim that it is a scientific discovery and that it is proven to be correct. I just want to share my observations with others and show them what I have seen and learnt.

It is very important to me that all the photos in the book portray real life situations. We never set dogs up to react and often had to forego some potentially fascinating footage in order to de-escalate the situation and keep everybody safe..

I hope that this book will reach many dog guardians and will help them to better understand their dogs' emotions. Understanding creates sympathy and compassion, two important pillars to foster a good relationship.

Katja Krauss runs the prestigious GREH dog school in Berlin. She is also among a select group of Tellington TTouch Instructors for dogs and small animals.

THE AUTHORS

GABI MAUE

How this book came to life

I have been drawn to animals, especially dogs and horses, for as long as I can remember. My parents were not able to fulfill my dream of having my own horse or dog when I was a child and I was almost 20 before a dog finally joined my life.

When Tiber, a 12-week old Pyrenean Shepherd, moved in with me, my passion for dogs turned into a deep and intense relationship. Tiber lost his sight when he was only three years old and he became very dependent on his humans. I learnt to read the other dogs for him, to predict their moods and reactions, so I could assure him that his interactions were safe. To make Tiber's life easier, we took on another dog of the same breed and age, and Fanny joined our family. When she first came to us she was very insecure when initially meeting unfamiliar people or dogs of a breed other than her own.

Unfortunately she was quick to go on the offensive, and snapped at dogs easily. Fanny taught me to read her signals, to see and respect her calls for help, and do so before she had to escalate her reactions. Through Fanny, I realised how subtly and quickly moods and emotions are conveyed, and how quickly they can change. With help and support, Fanny became a confident and trusting dog. Sadly, we lost her at the age of nine. Tiber did not deal well with Fanny's death, so we quickly looked for a new puppy, and that was when Quivive, an 11-week-old puppy, joined the family.

Quivive was raised with Tellington TTouch® and our relationship was based on mutual trust and respect. She became my partner and co-teacher in my workshops, group sessions, and one-to-one consultations. Quivive became a master in reading, and using, body language and signals when communicating with other dogs. I learnt something new from her every single day.

During my training to become a Tellington TTouch® practitioner, I met my fellow author, Katja Krauss. I was fascinated, not only with her skill in reading and evaluating dogs, but also how quickly she could make good decisions for dogs and humans, and then execute them. While working with Katja, the idea for this book was hatched. We started collecting photos and soon we were searching through literally thousands of photos to find exactly what we wanted. It has taken many years for this project to come to fruition and we are thrilled with the finished result. It has been a journey of hard work, love and dedication. Our abiding hope is that readers will learn from it and, as a result of their new understanding, our dogs will live better, happier lives.

Gabi Maue is a Level 3 Tellington TTouch practitioner for dogs and small animals. She hosts seminars and lectures on the Tellington TTouch method, on communication, learning behavior and behavioral problems as well as on the neuropsychology of dogs.